STUDENT LECTURE NOTEBOOK

Plant Biology

SECOND EDITION

GRAHAM ▪ GRAHAM ▪ WILCOX

University of Wisconsin

PEARSON
Prentice Hall

Upper Saddle River, NJ 07458

Publisher: Sheri L. Snavely
Project Manager: Karen Horton
Executive Managing Editor: Kathleen Schiaparelli
Assistant Managing Editor: Becca Richter
Production Editor: Gina M. Cheselka
Supplement Cover Manager: Paul Gourhan
Supplement Cover Designer: Joanne Alexandris
Manufacturing Buyer: Ilene Kahn
Cover Photo Credit: Lee W. Wilcox

Pearson Prentice Hall
Pearson Education, Inc.
Upper Saddle River, NJ 07458

Printed in the United States of America

10 9 8 7 6 5 4 3 2 1

ISBN 0-13-146908-8

Pearson Education Ltd., *London*
Pearson Education Australia Pty. Ltd., *Sydney*
Pearson Education Singapore, Pte. Ltd.
Pearson Education North Asia Ltd., *Hong Kong*
Pearson Education Canada, Inc., *Toronto*
Pearson Educación de Mexico, S.A. de C.V.
Pearson Education—Japan, *Tokyo*
Pearson Education Malaysia, Pte. Ltd.

Contents

CHAPTER 1 Introduction to Plant Biology

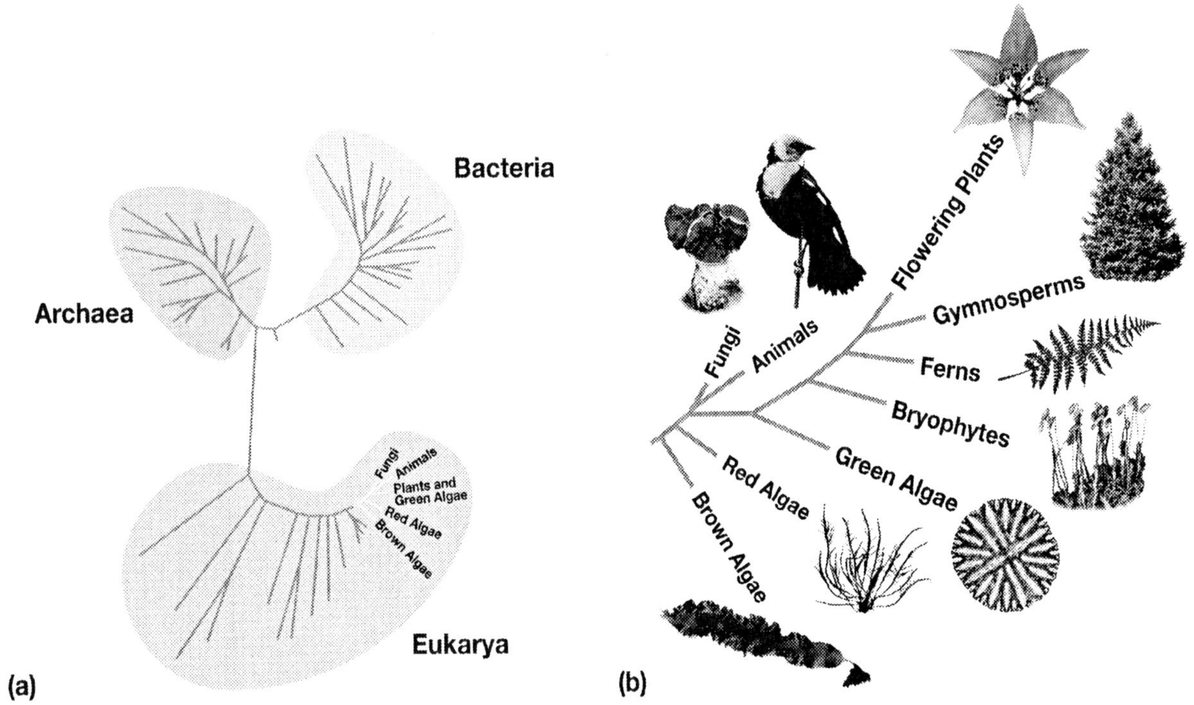

Figure 1.5 **A Tree of life shows current views of the relationships among plants, algae, fungi, bacteria, and other organisms**

NOTES

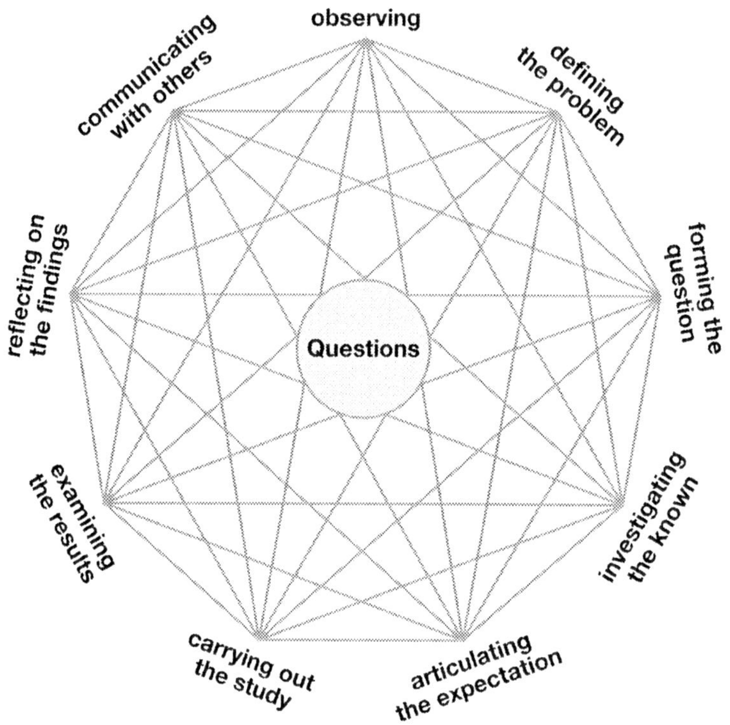

Figure 1.7 **Processes involved in the scientific method**

NOTES

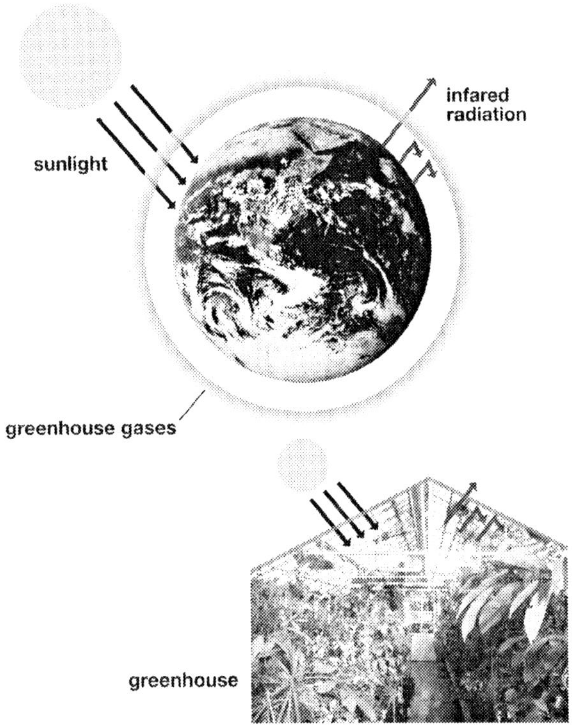

Figure 1.8 **The greenhouse effect**

NOTES

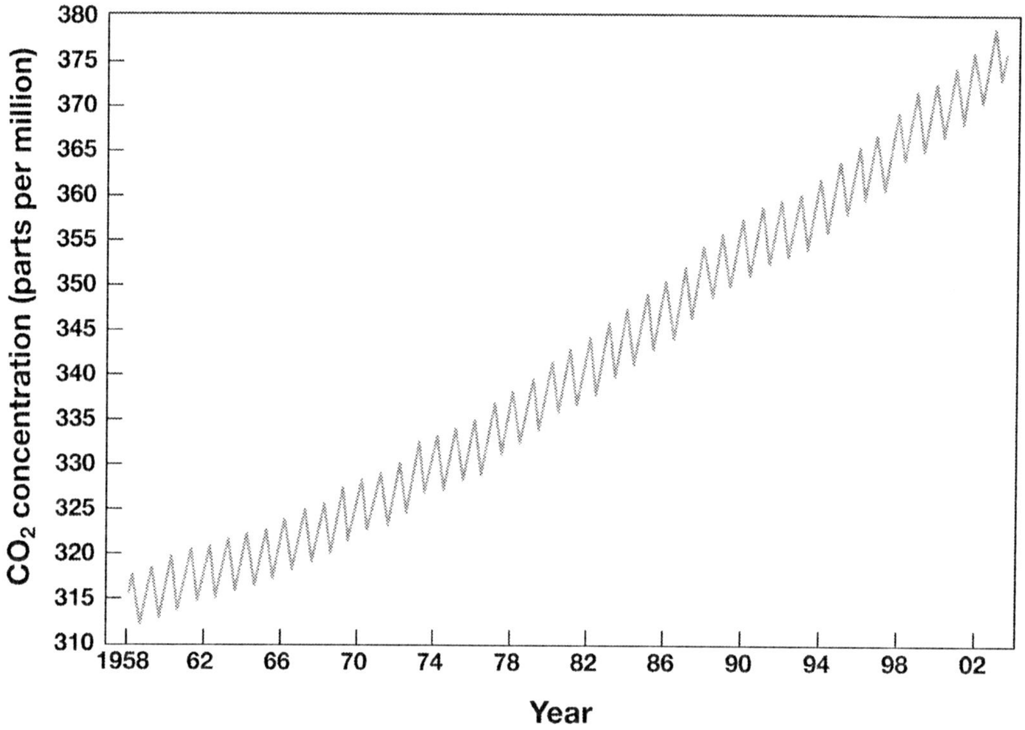

E1.2 **Graph showing atmospheric carbon dioxide increases since the late 1950s as measured in Hawaii**

NOTES

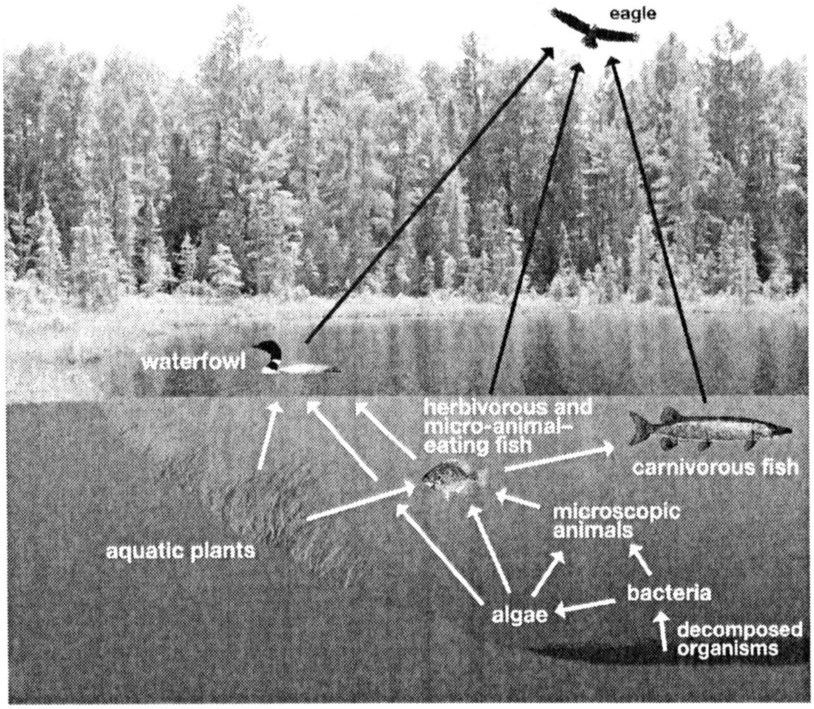

Figure 1.9 **Food Web**

NOTES

NOTES

CHAPTER 3

Molecules of Life

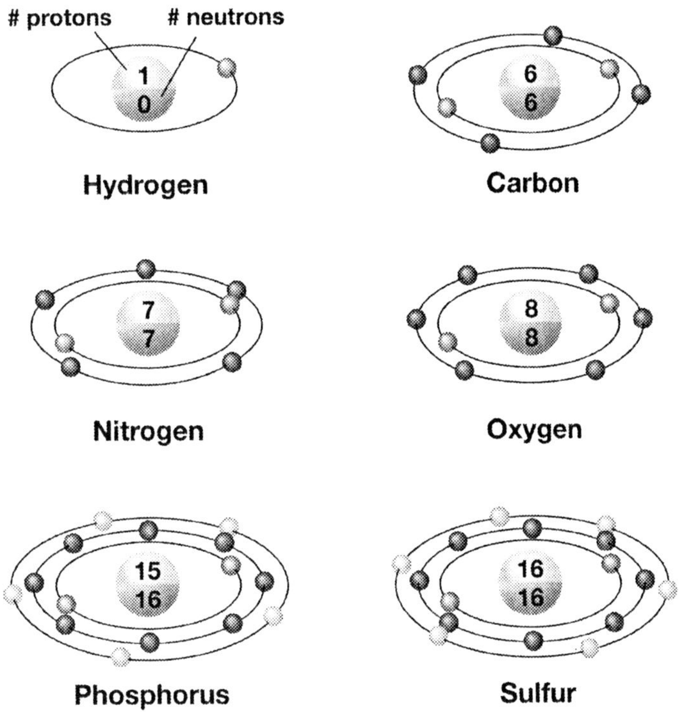

Figure 3.1 **The six most common elements in plants**

NOTES

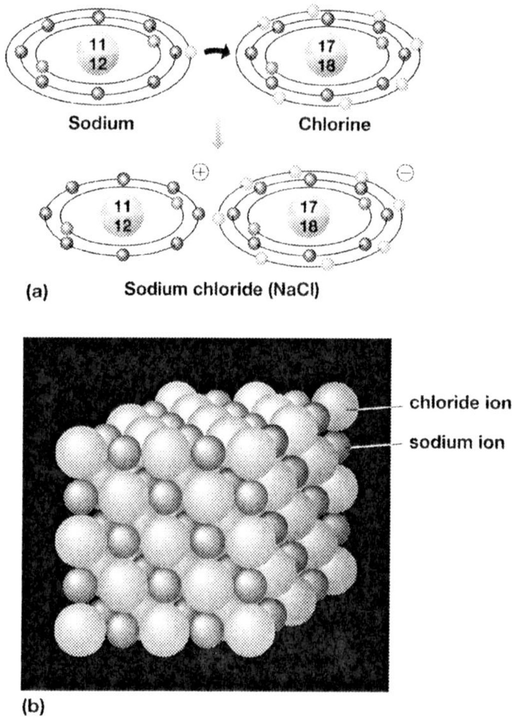

Figure 3.4 **Diagram of ionic bonds between atoms of sodium and chlorine**

NOTES

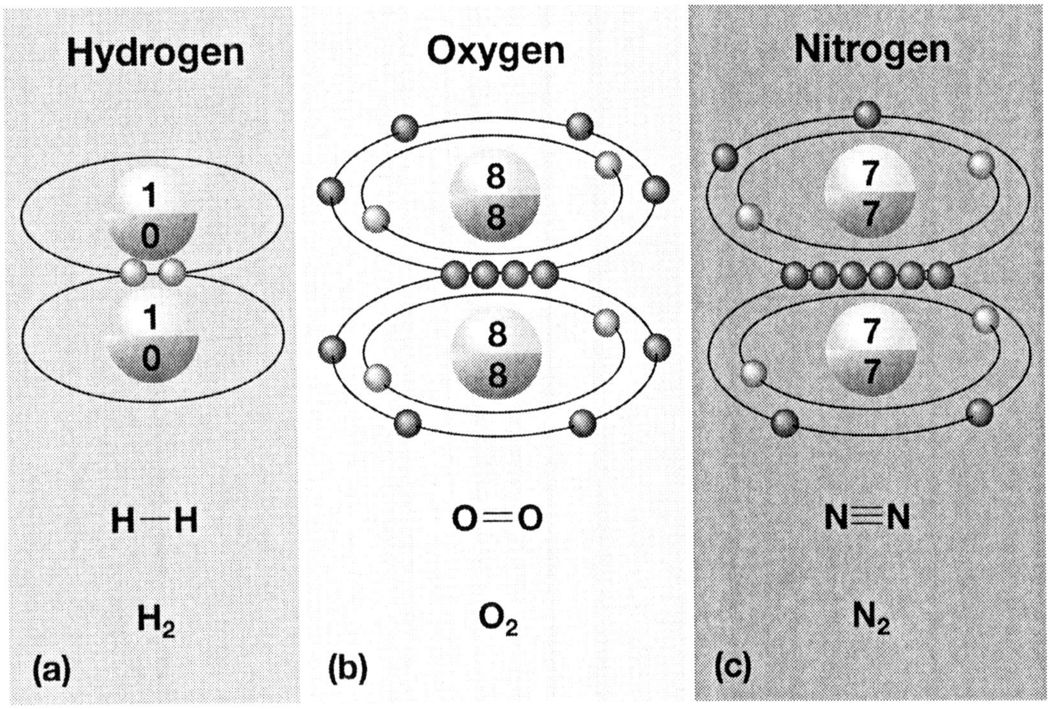

Figure 3.5 **Covalent bonds in molecular (a) hydrogen, (b) oxygen, and (c) nitrogen**

NOTES

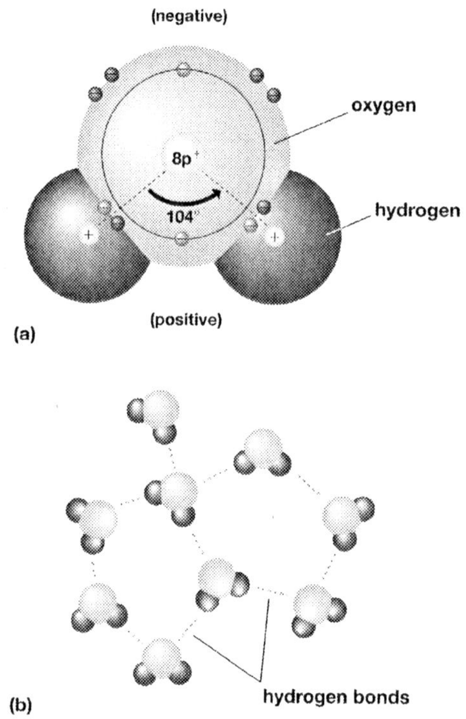

Figure 3.6 **Water**

NOTES

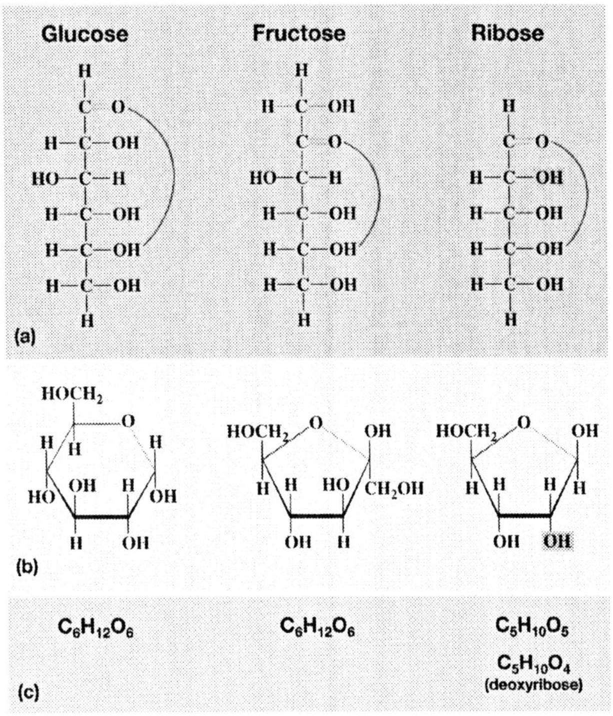

Figure 3.8 **Diagrams of the molecular structure of some sugars that occur in plants**

NOTES

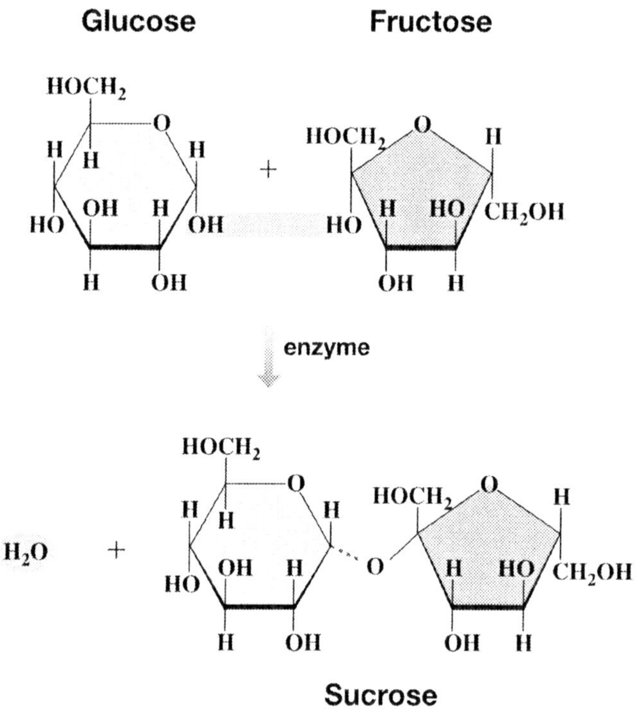

Figure 3.9 **Diagram of dehydration synthesis of sucrose from glucose and fructose**

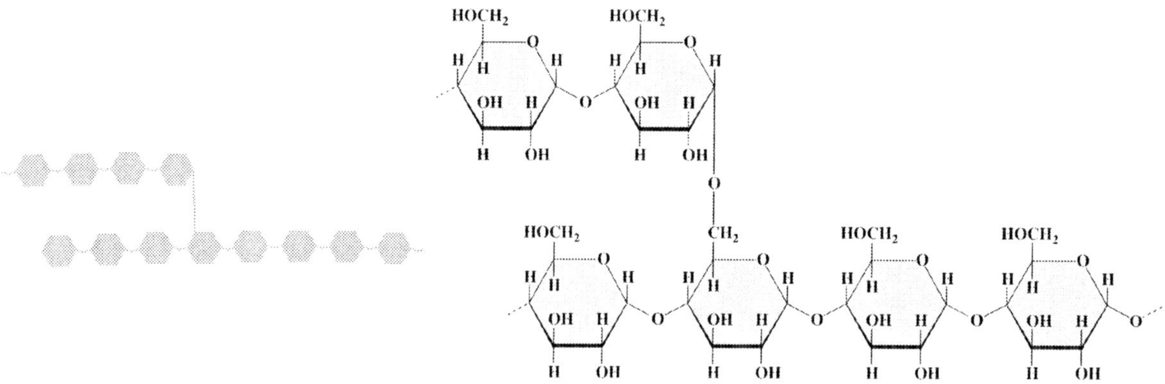

Figure 3.10 **Starch as an energy storage molecule in plants**

NOTES

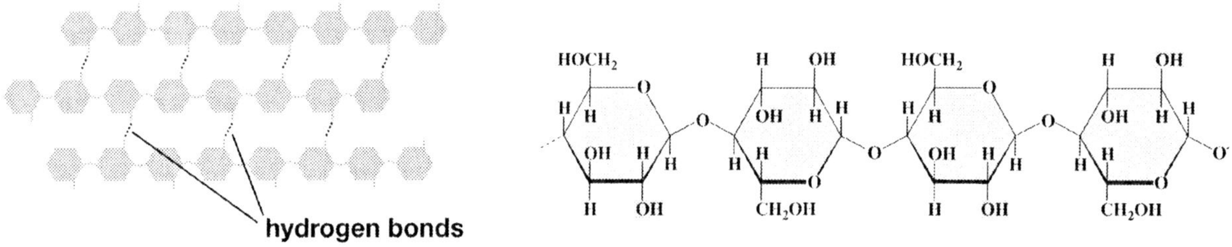

Figure 3.11 **Cellulose is a carbohydrate that provides plant structure**

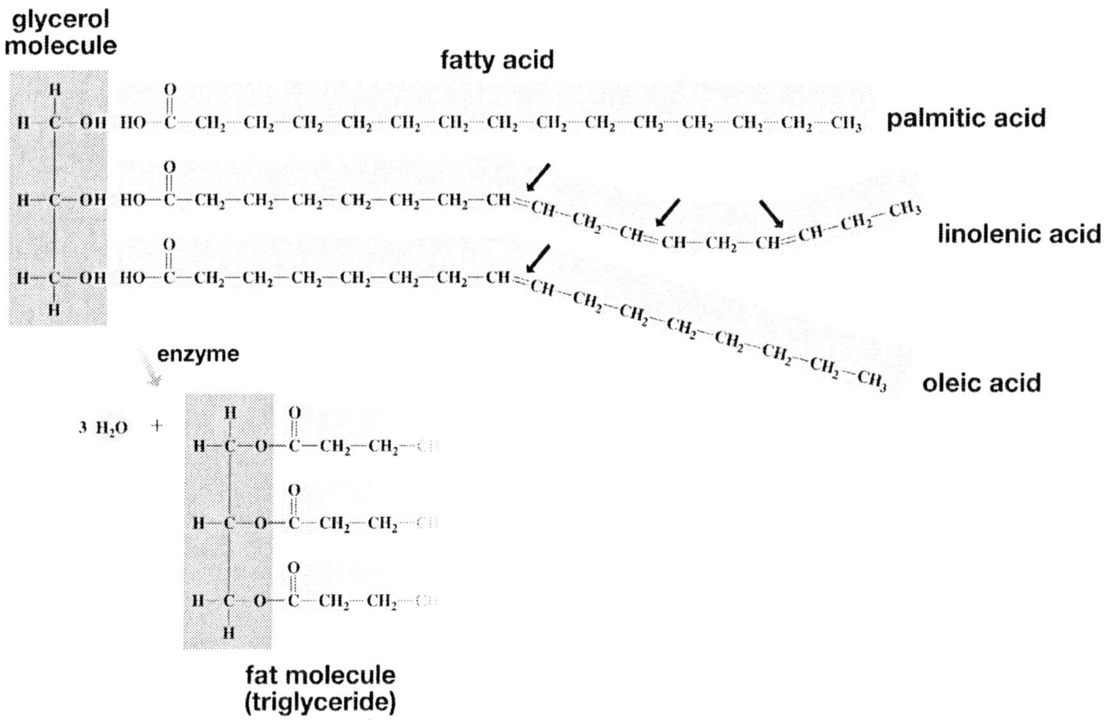

Figure 3.12 **Diagram showing the formation of a fat molecule (triglyceride), which consists of glycerol plus three fatty acids**

NOTES

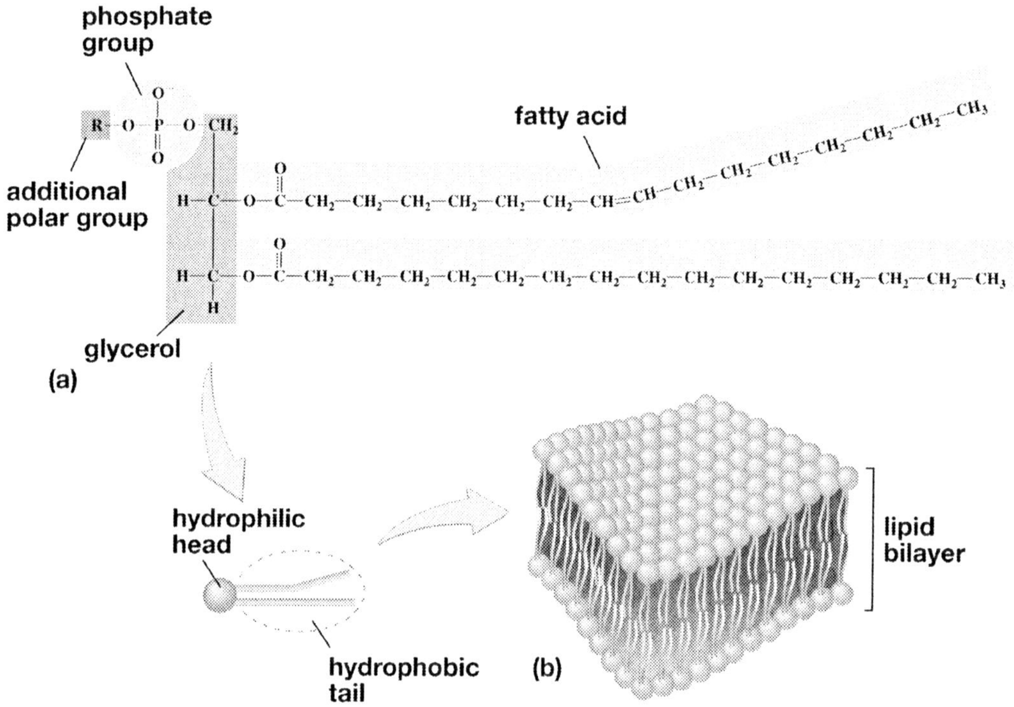

Figure 3.14 **Phospholipids**

NOTES

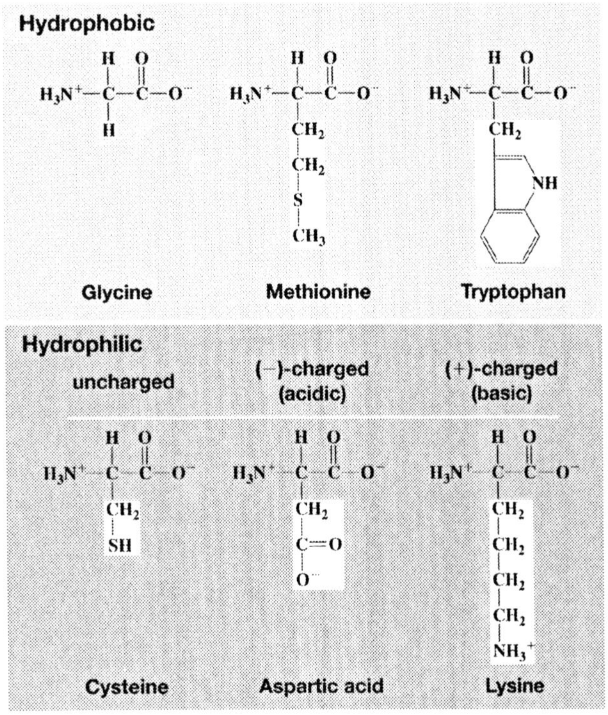

Figure 3.16 **Amino acids**

NOTES

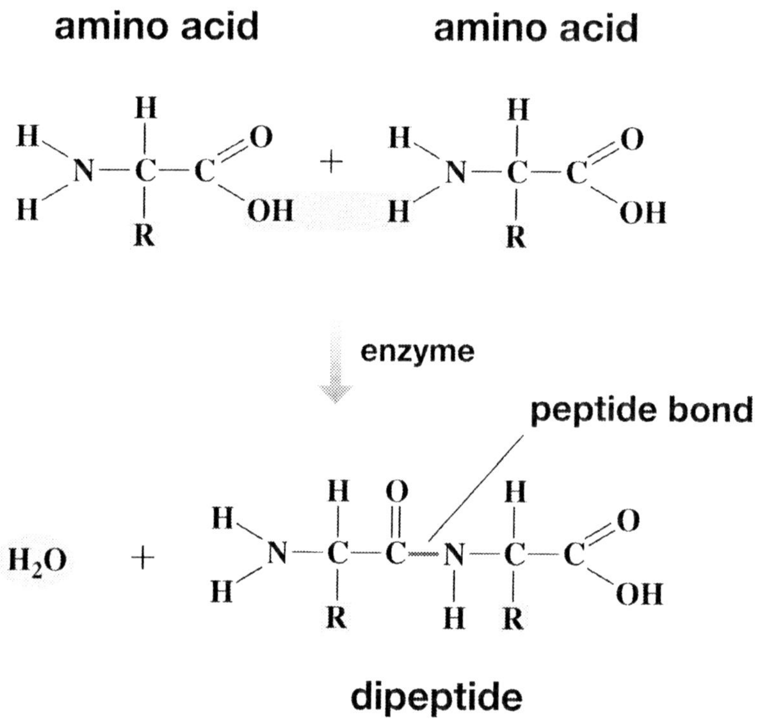

Figure 3.17 **Diagram showing the formation of a peptide bond**

NOTES

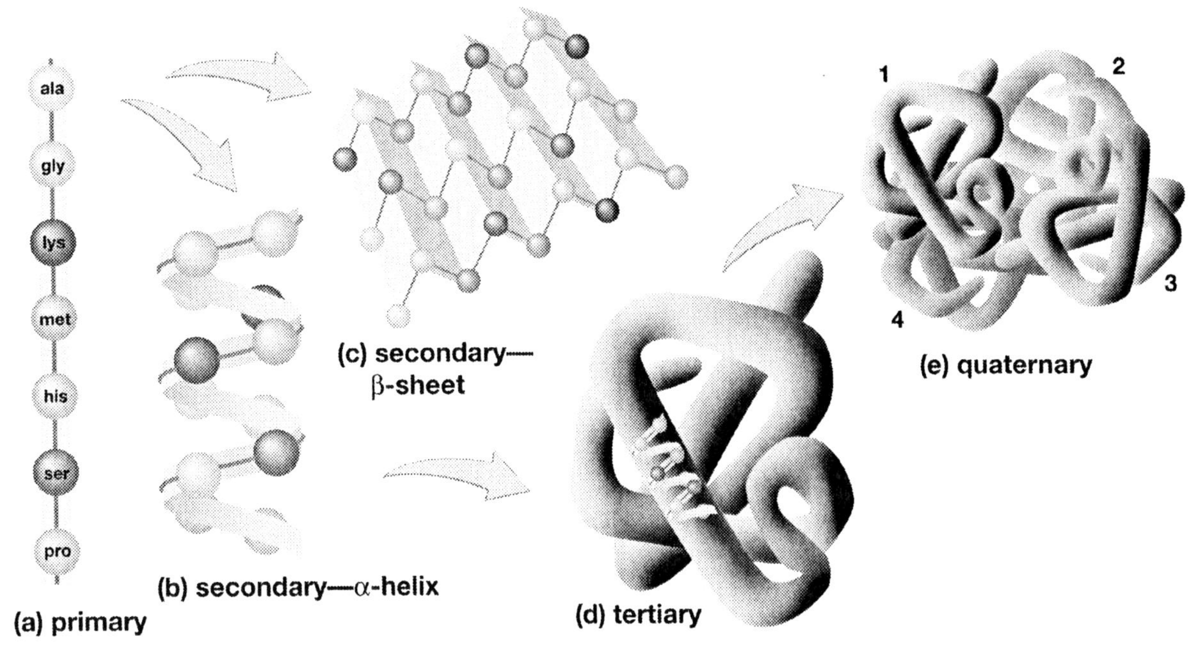

Figure 3.18 **Hierarchy of protein structure**

NOTES

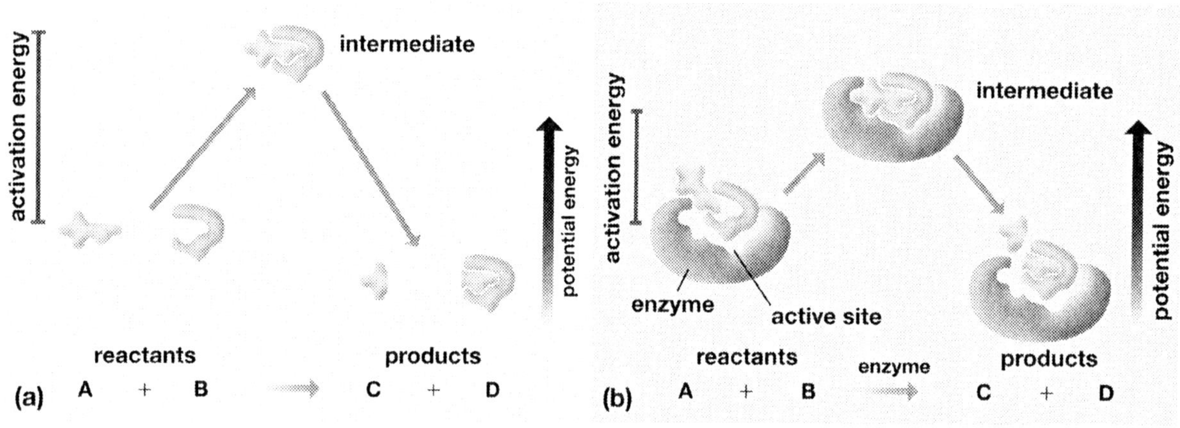

Figure 3.19 **A chemical reaction and an enzyme-catalyzed biochemical reaction**

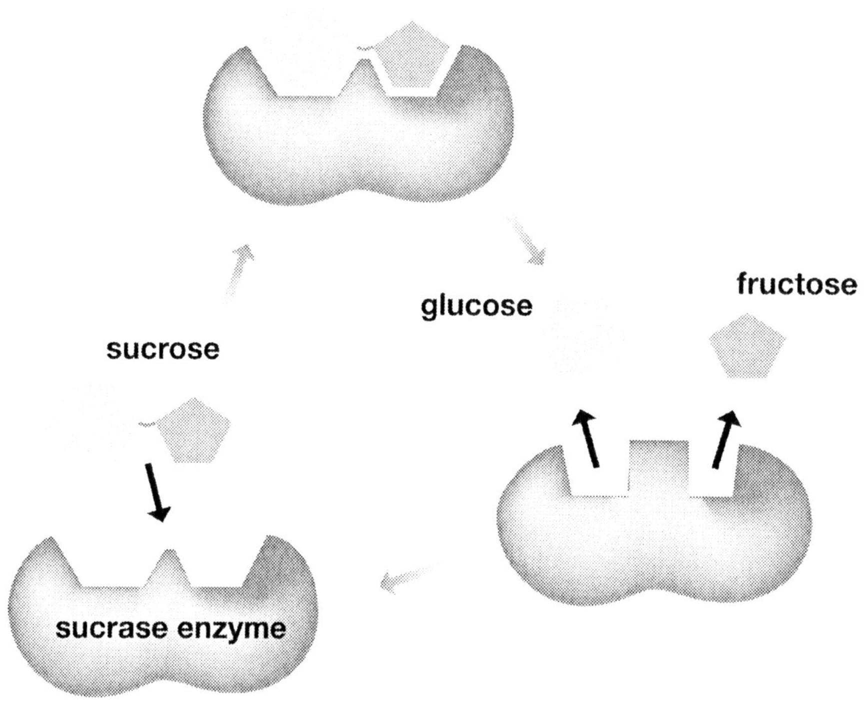

Figure 3.20 **Diagram of the enzyme sucrase (invertase) breaking sucrose into glucose and fructose**

NOTES

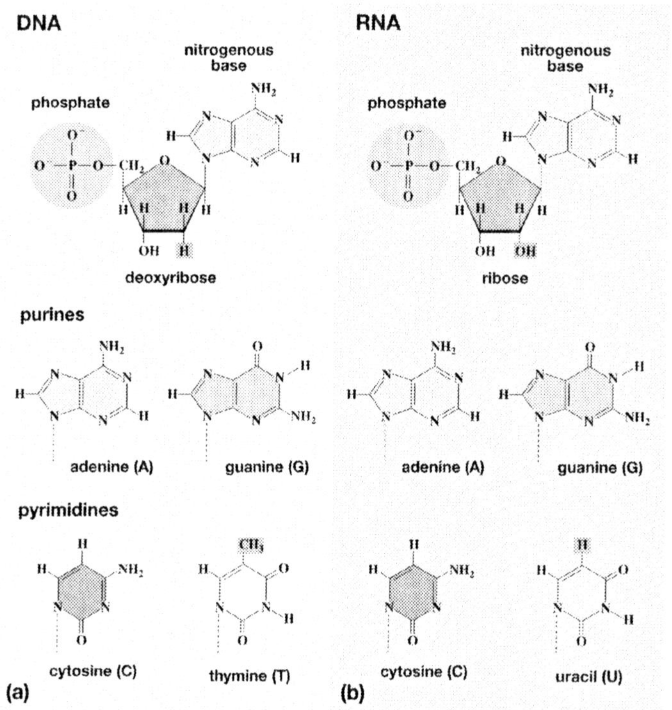

Figure 3.21 **Nucleotides and nitrogenous bases in DNA and RNA**

NOTES

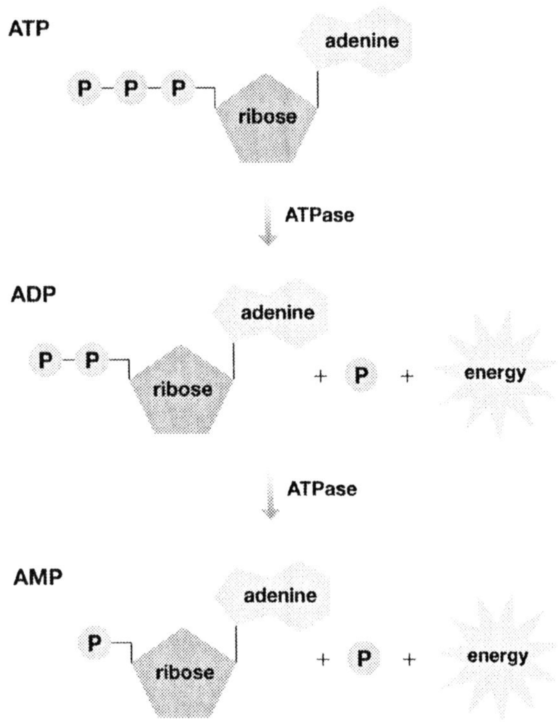

Figure 3.23 **Adenosine triphosphate (ATP)**

NOTES

NOTES

CHAPTER 4 | Cells

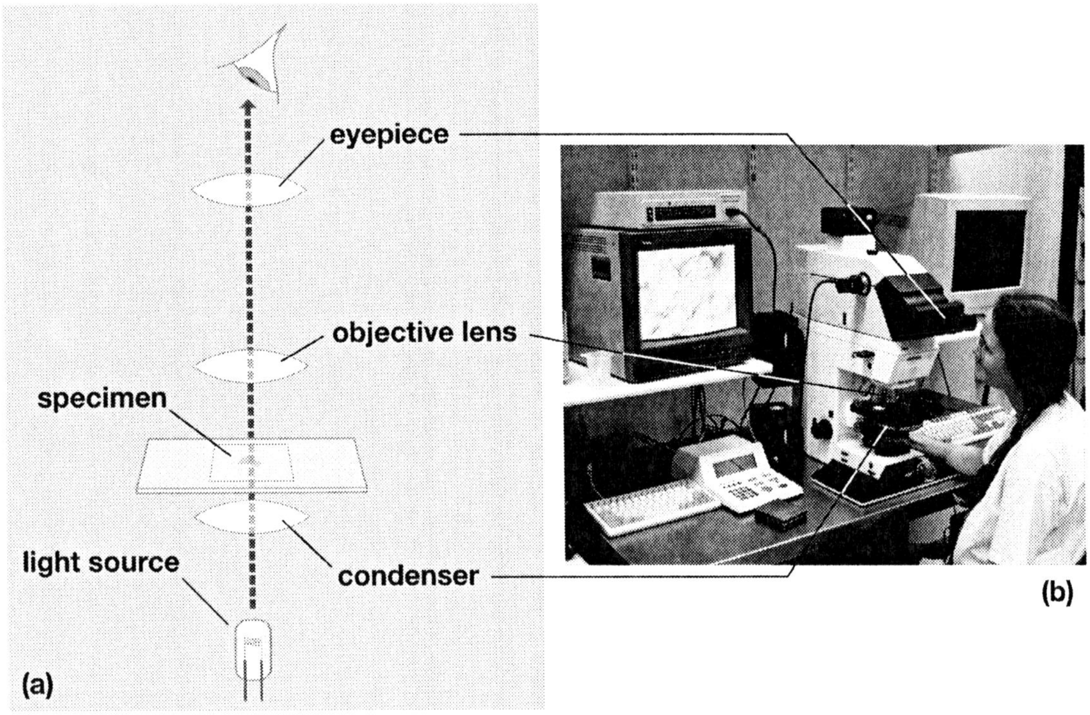

Figure 4.2 **Compound light microscopy**

NOTES

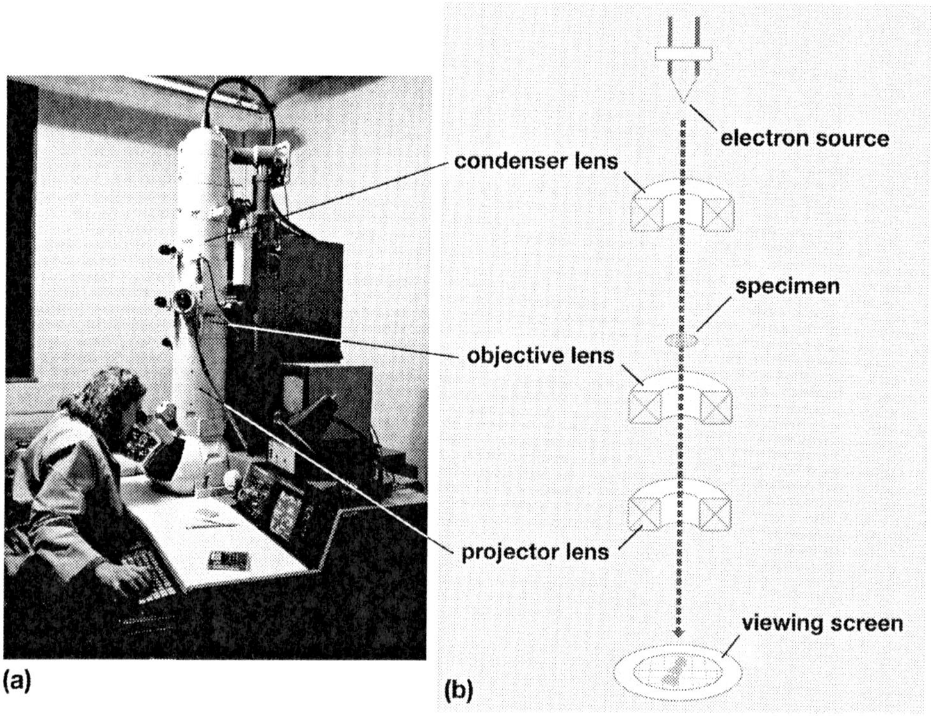

Figure 4.6 **Transmission electron microscopy**

NOTES

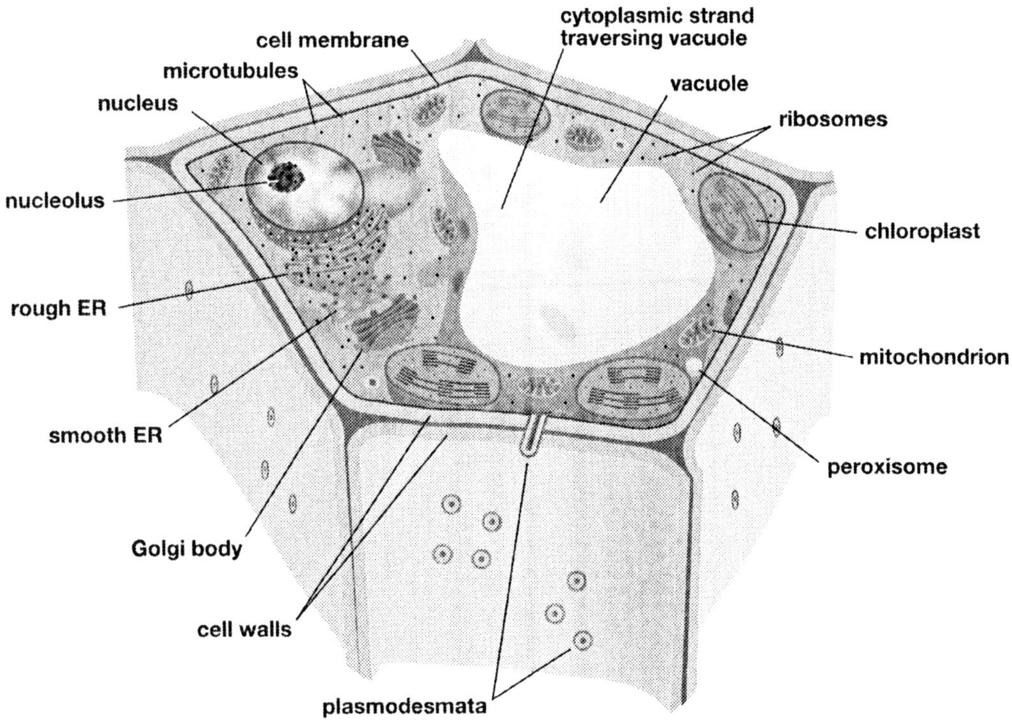

Figure 4.7 **Diagram of a plant cell**

NOTES

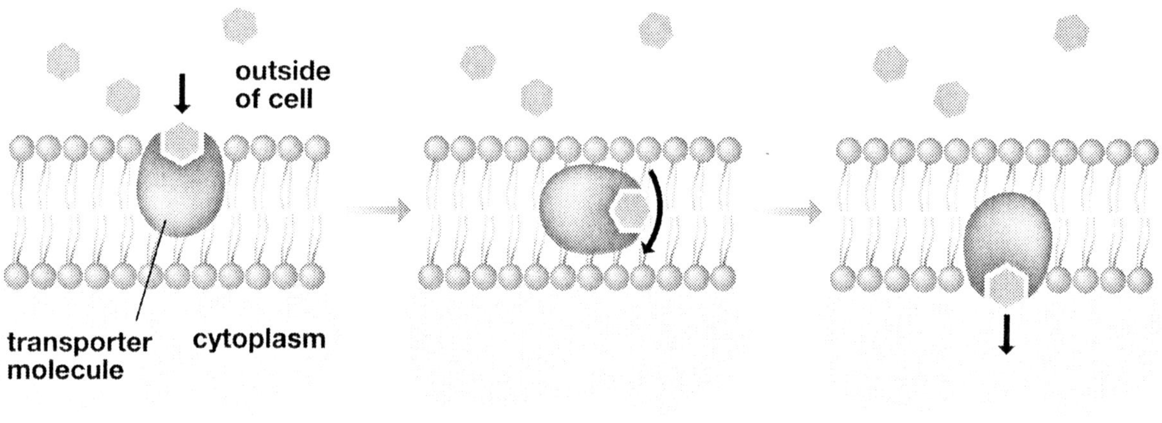

Figure 4.9 **Cell-membrane carrier protein**

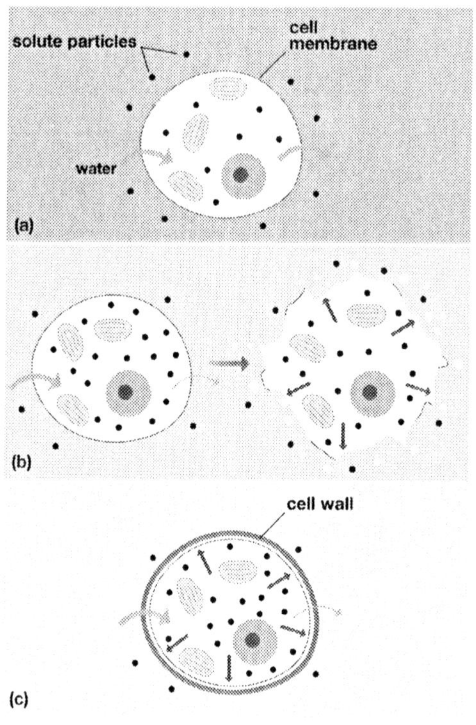

Figure 4.10 **Solute balance**

NOTES

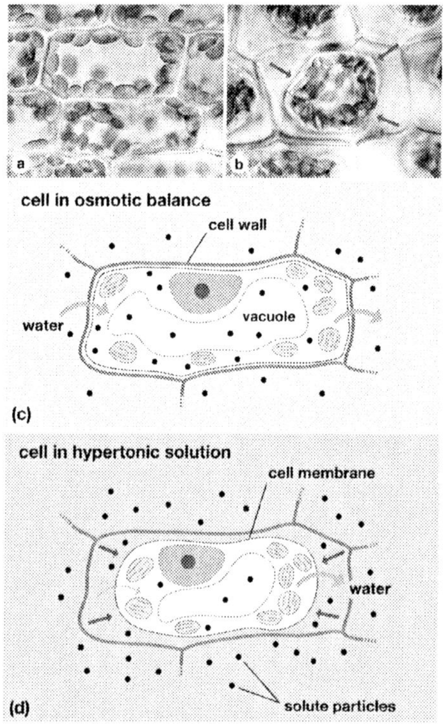

Figure 4.12 **Plasmolysis**

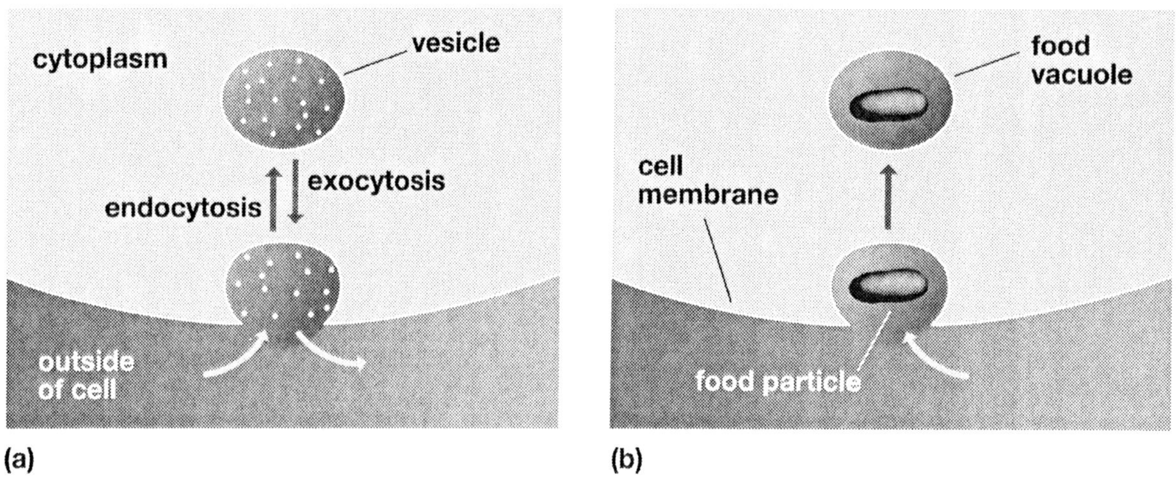

Figure 4.15 **Exocytosis, endocytosis, and phagocytosis**

NOTES

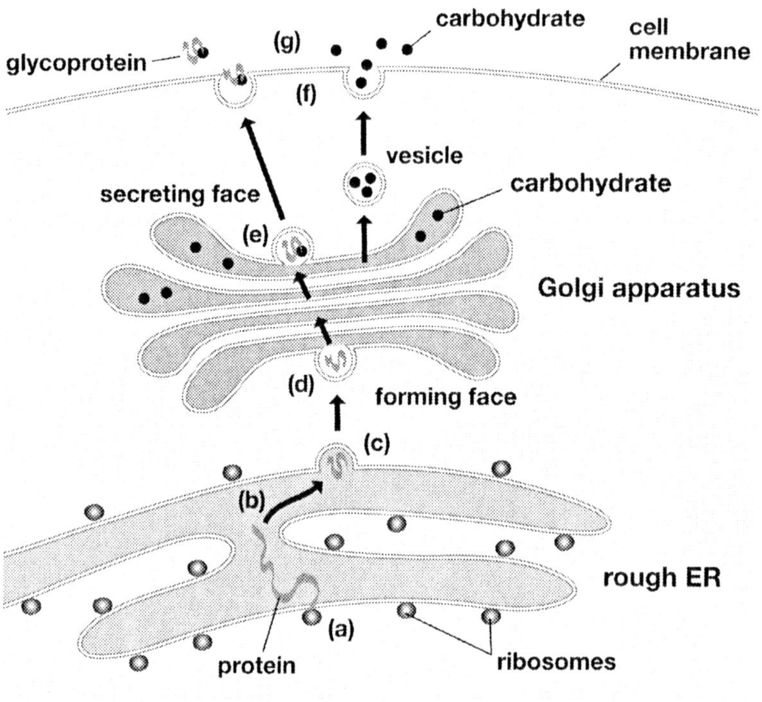

Figure 4.19 **Endomembrane system**

NOTES

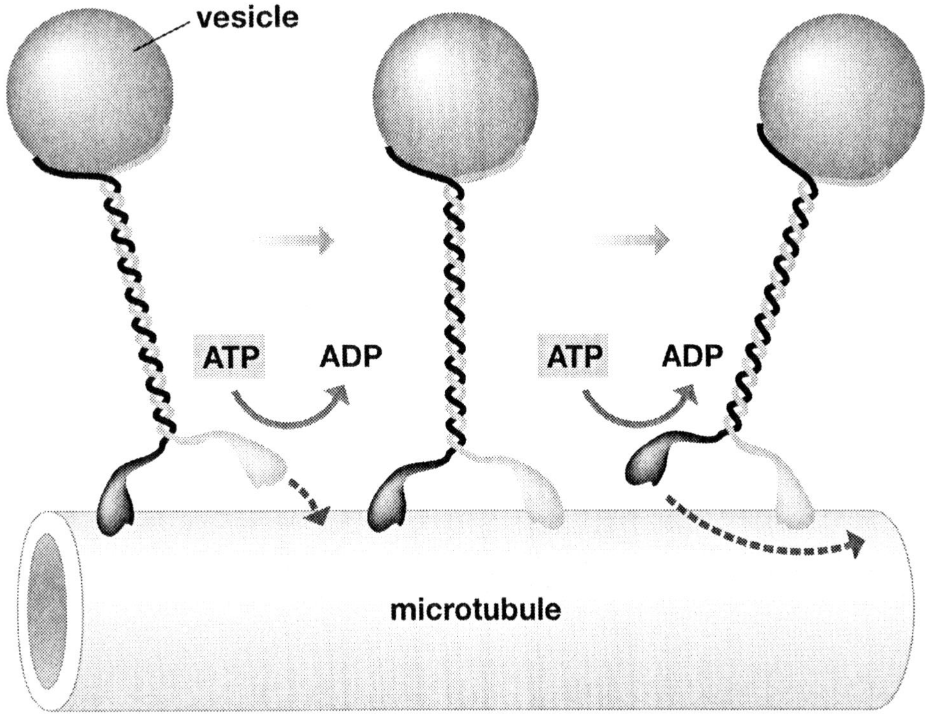

Figure 4.20 **The motor protein kinesin moves vesicles along microtubule tracks**

NOTES

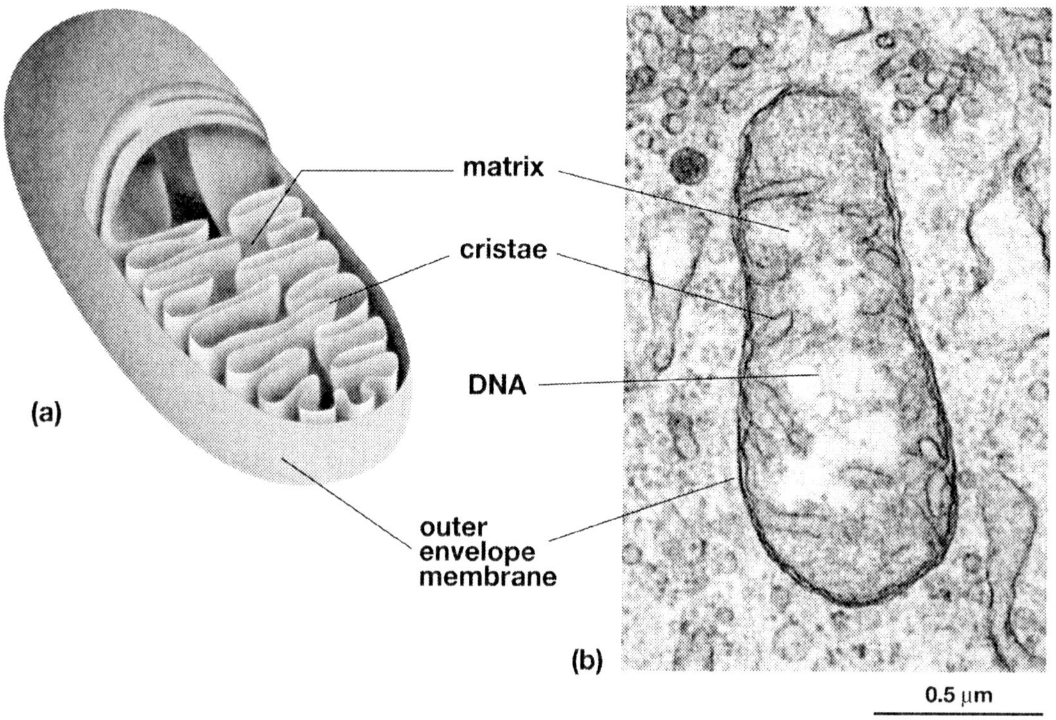

Figure 4.22 **Mitochondrion**

NOTES

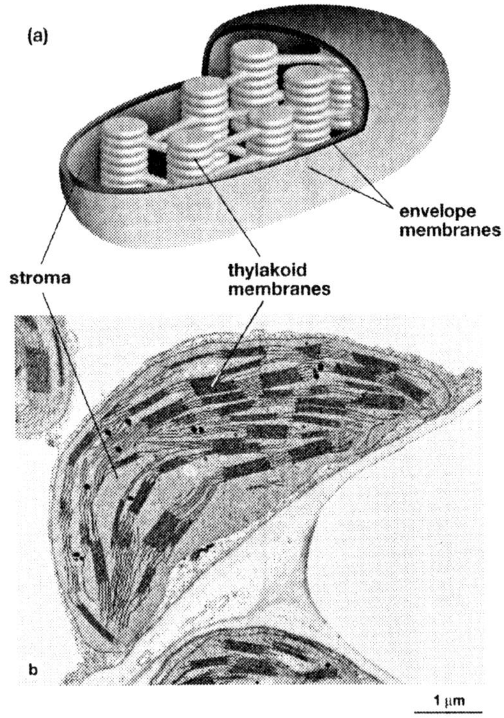

Figure 4.27 **Chloroplast**

NOTES

NOTES

CHAPTER 5

Photosynthesis and Respiration

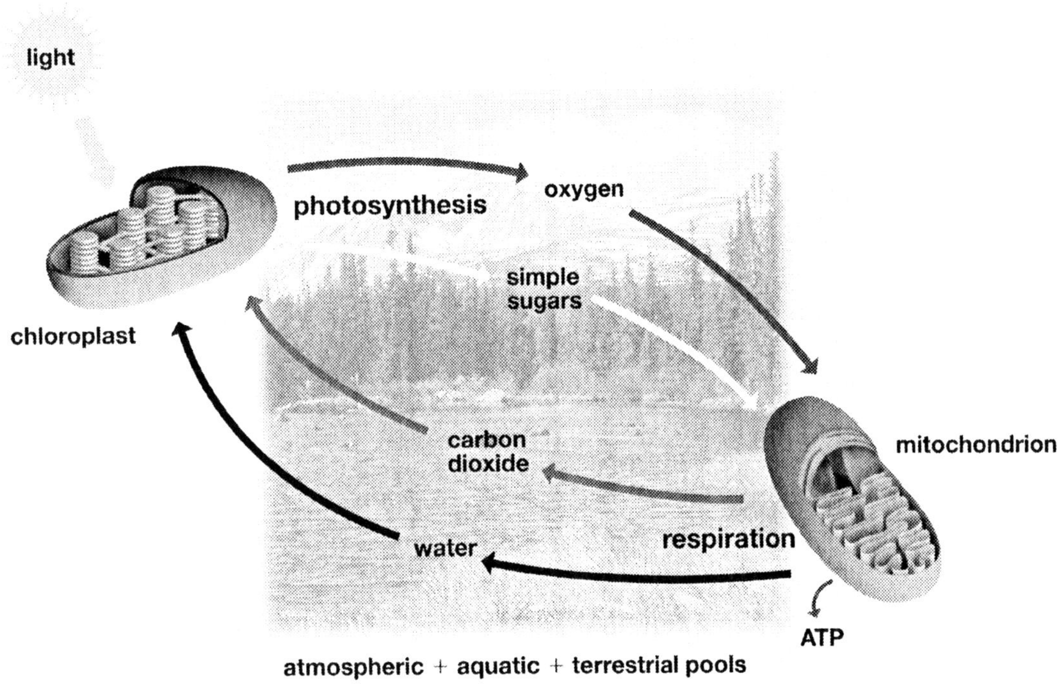

Figure 5.1 **The path of energy flow on Earth**

NOTES

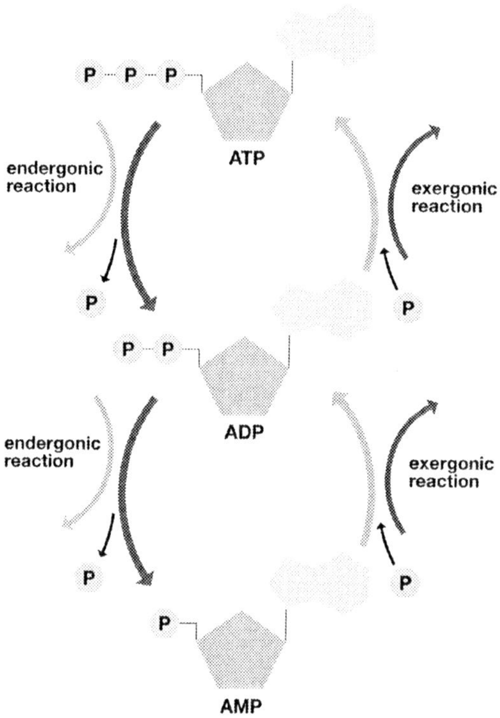

Figure 5.2 **ATP transfers energy between coupled reactions**

NOTES

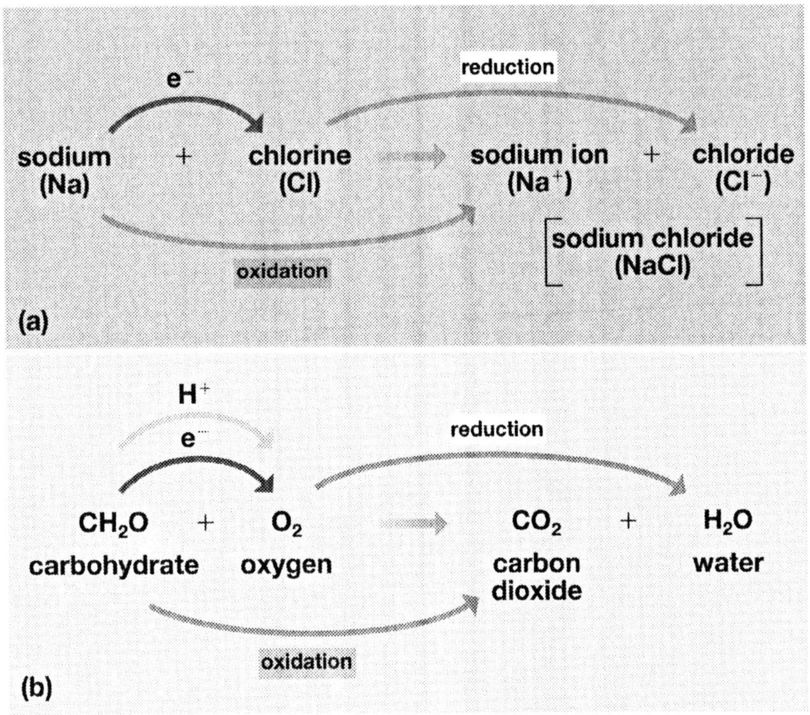

Figure 5.3 **Oxidation and reduction reactions**

NOTES

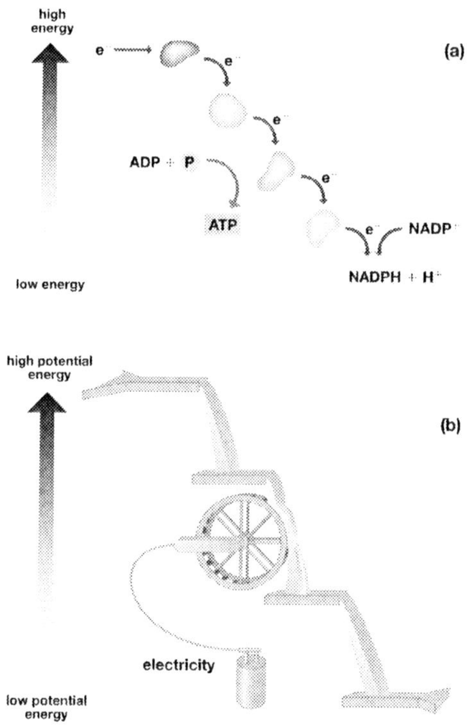

Figure 5.4 **An electron transport chain and a physical analogy**

NOTES

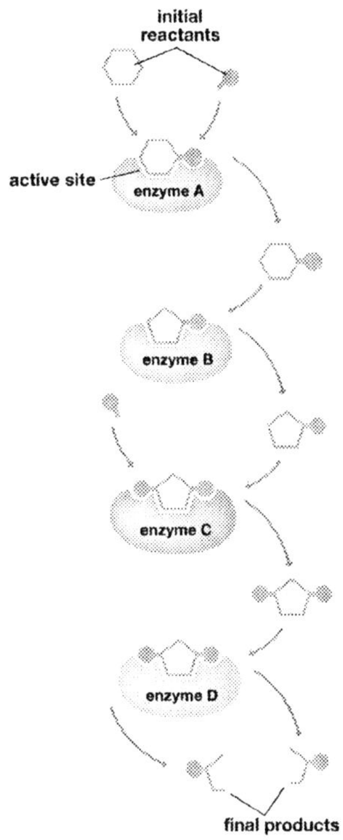

Figure 5.5 **An ordered series of chemical reactions**

NOTES

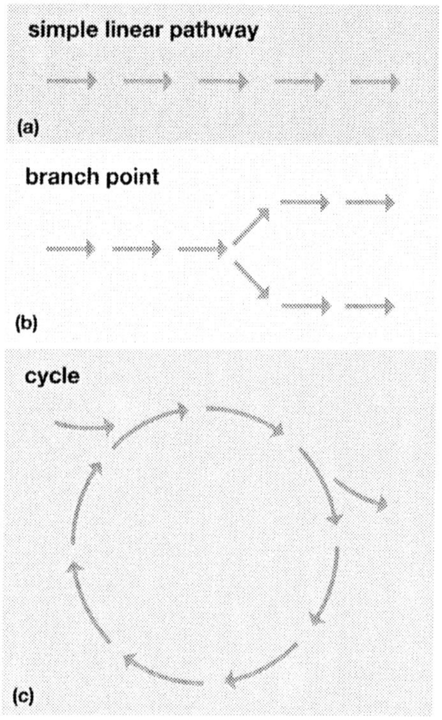

Figure 5.6 **Three basic types of metabolic pathways**

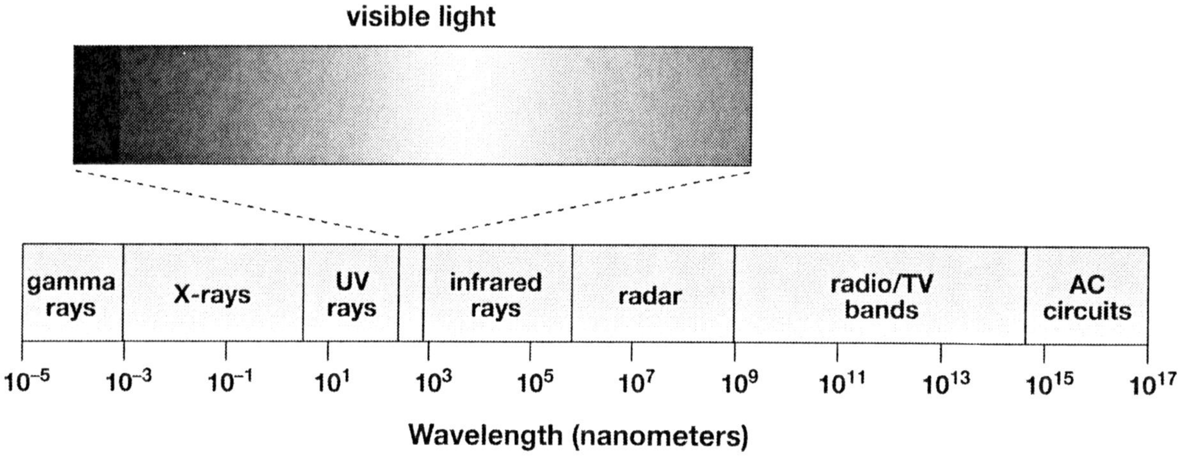

Figure 5.7 **The electromagnetic spectrum**

NOTES

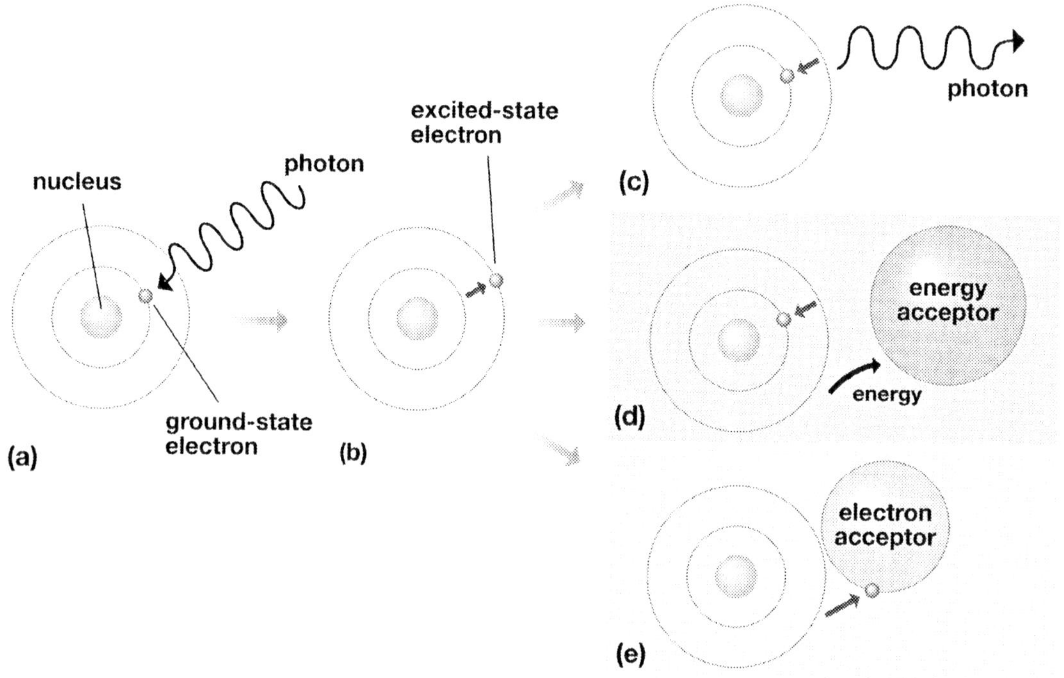

Figure 5.8 **The exitation of an electron by a photon of visible light**

NOTES

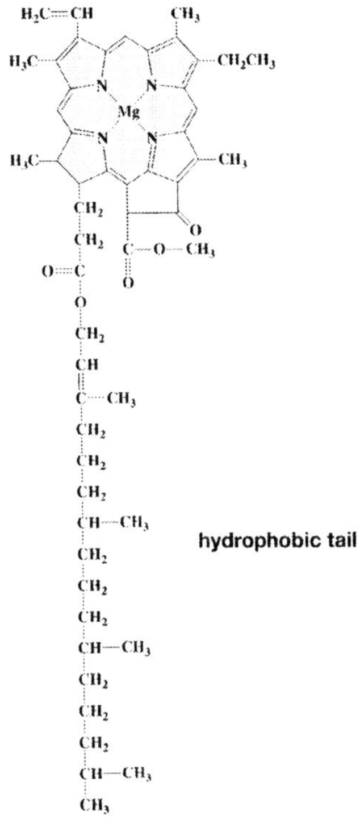

Figure 5.9 **The structure of chlorophyll *a***

NOTES

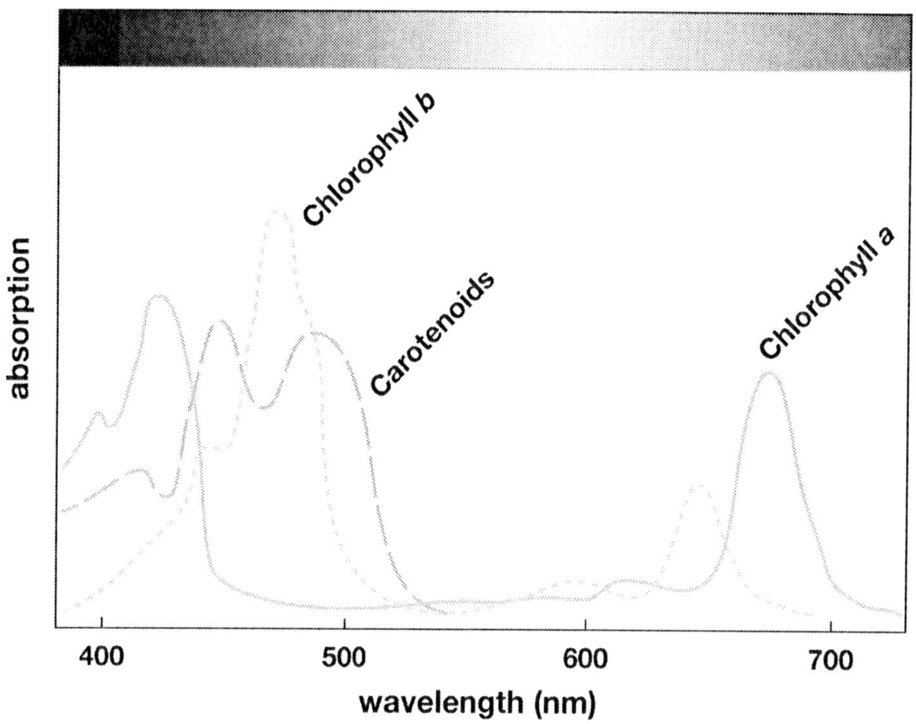

Figure 5.10 **Absorption spectrum of chlorophyll *a*, chlorophyll *b*, and carotenoids**

NOTES

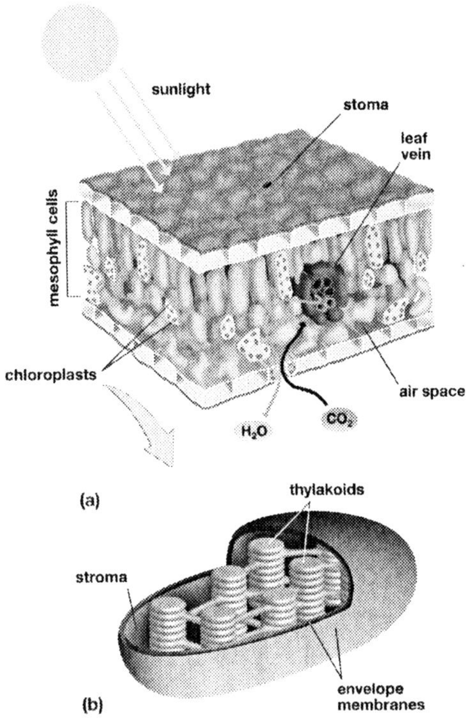

Figure 5.11 **Chloroplasts within leaf cells are the primary sites of photosynthesis**

NOTES

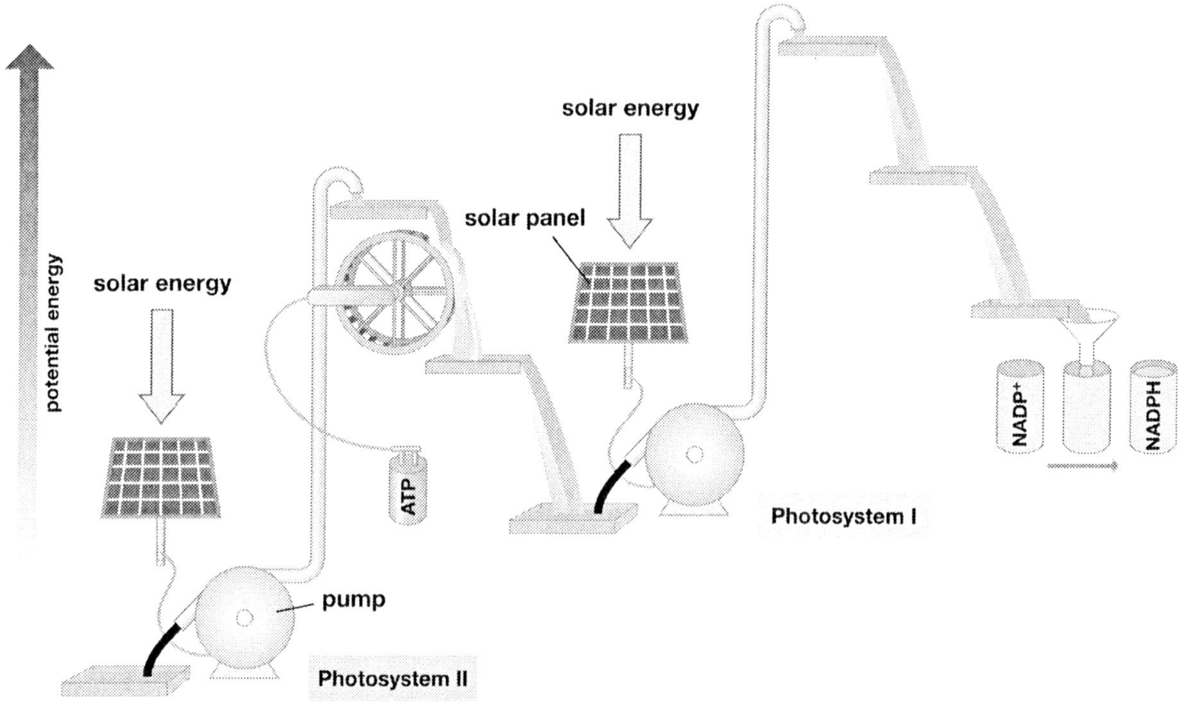

Figure 5.12 **Mechanical analogy to the light reactions in the chloroplast thylakoids**

NOTES

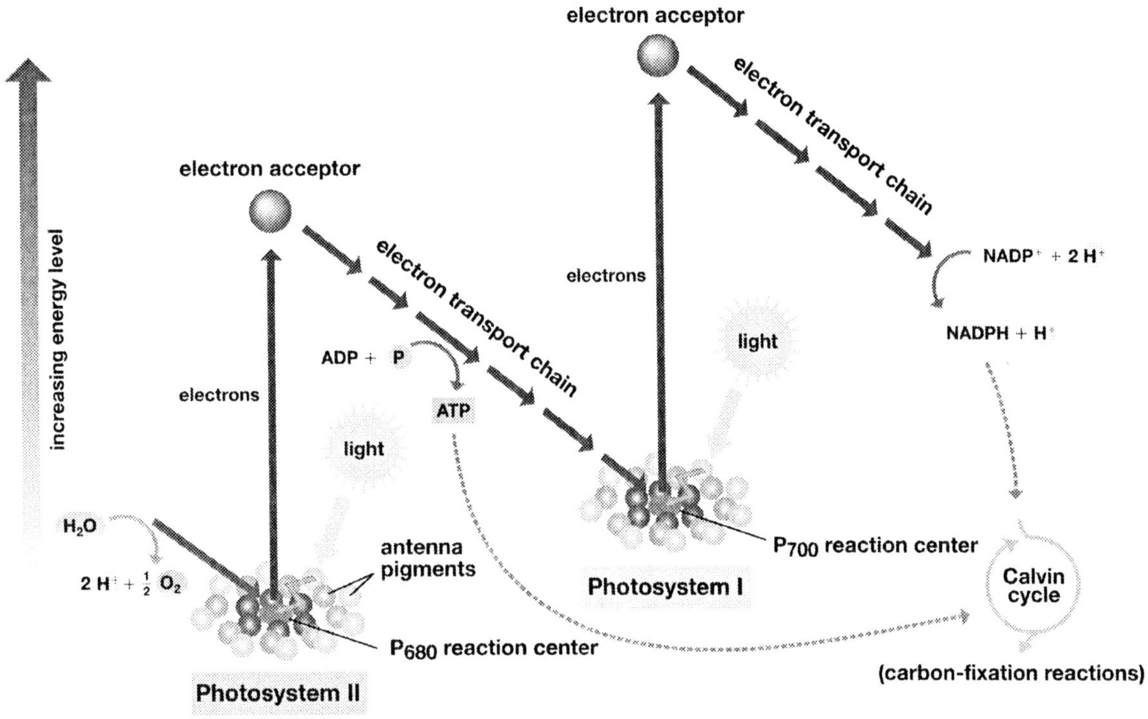

Figure 5.13 **Scematic diagram of the energy levels of electrons during the light reactions**

NOTES

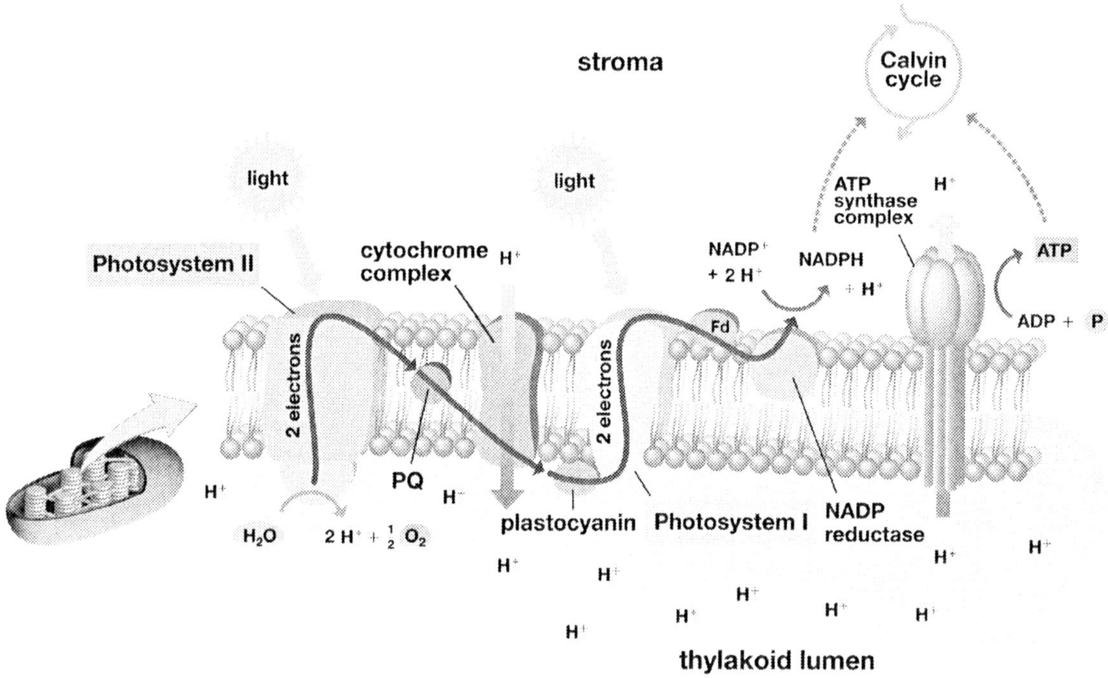

Figure 5.14 **The location of the molecules of the light reactions in the thylakoid membrane**

NOTES

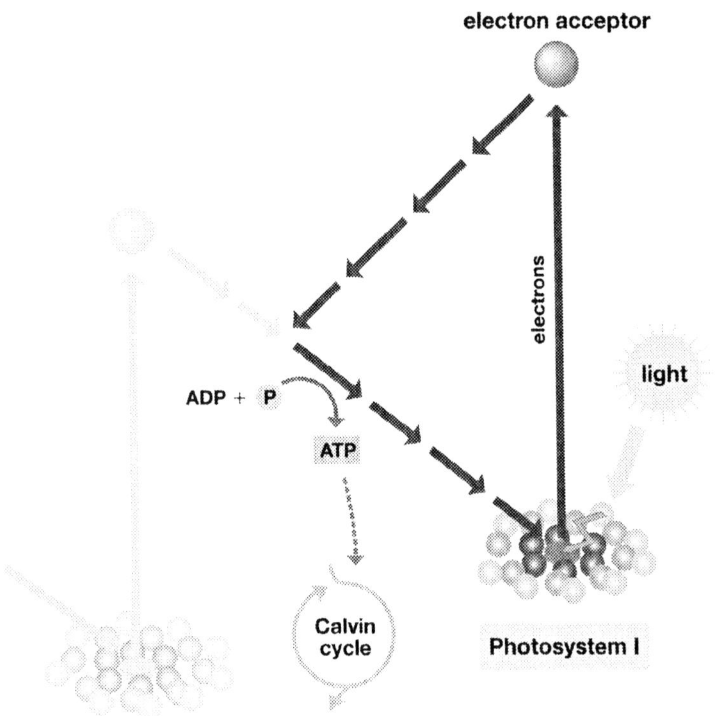

Figure 5.15 **Cyclic electron flow**

NOTES

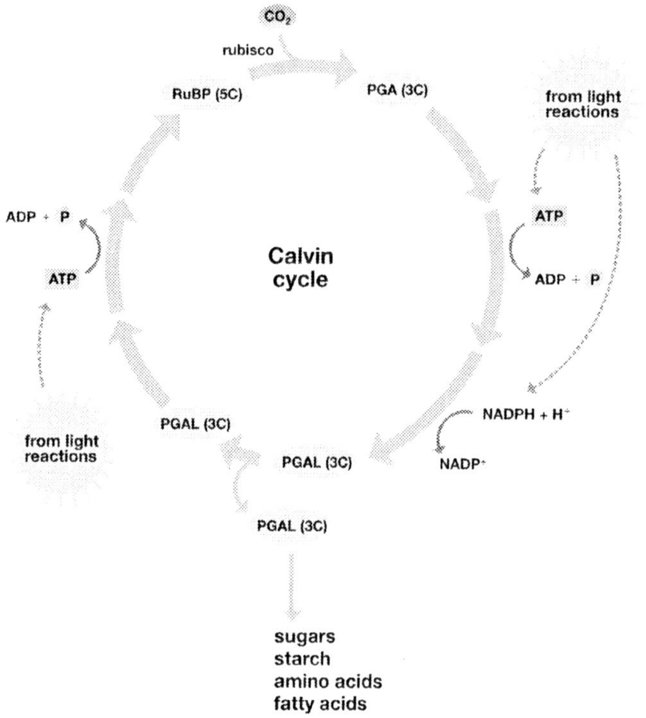

Figure 5.16 **Calvin cycle**

NOTES

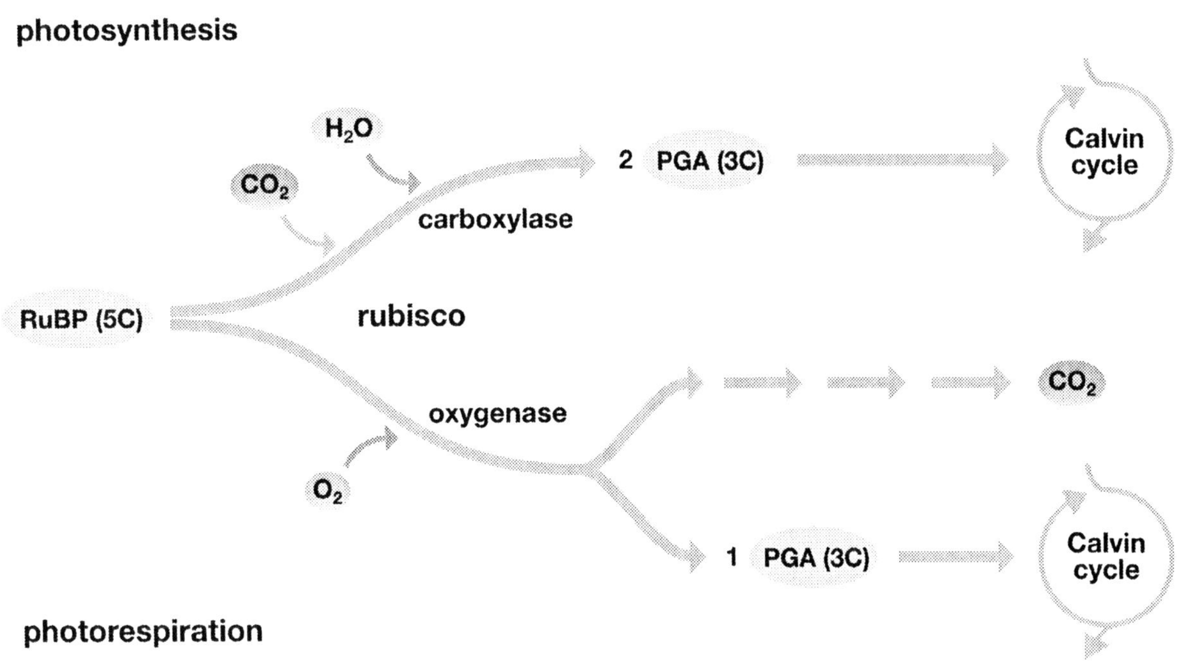

Figure 5.17 **Photorespiration**

NOTES

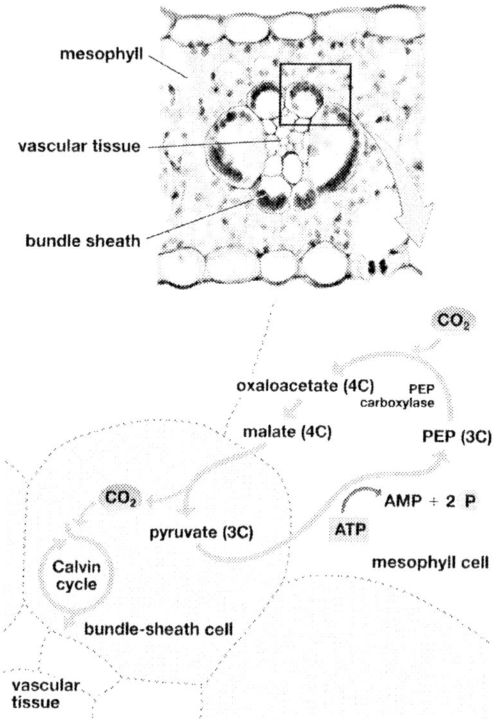

Figure 5.18 **C_4 pathway**

NOTES

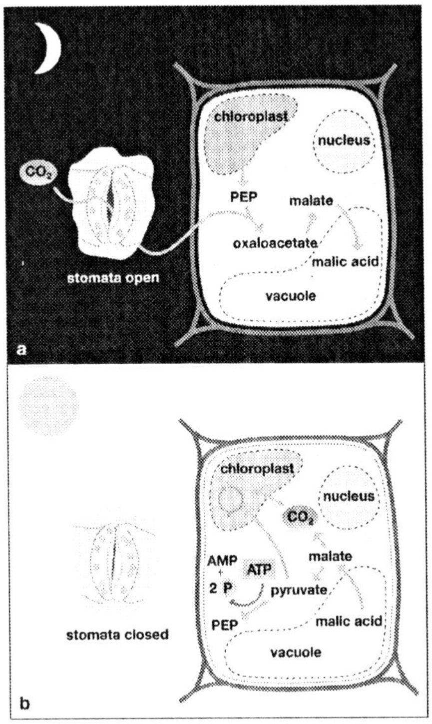

Figure 5.19 **Crassulacean acid metabolism**

NOTES

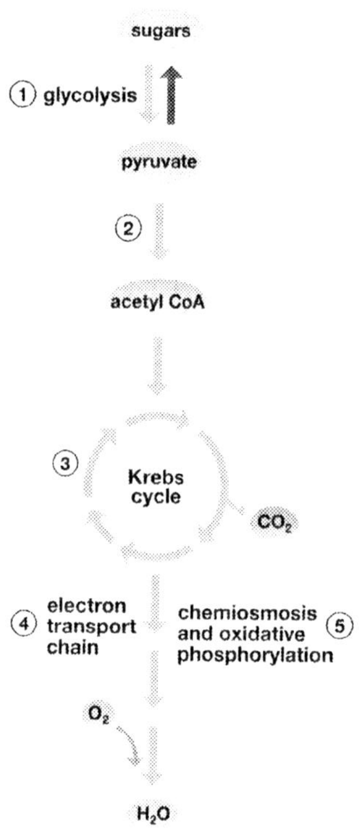

Figure 5.21 **The five processes in respiration**

NOTES

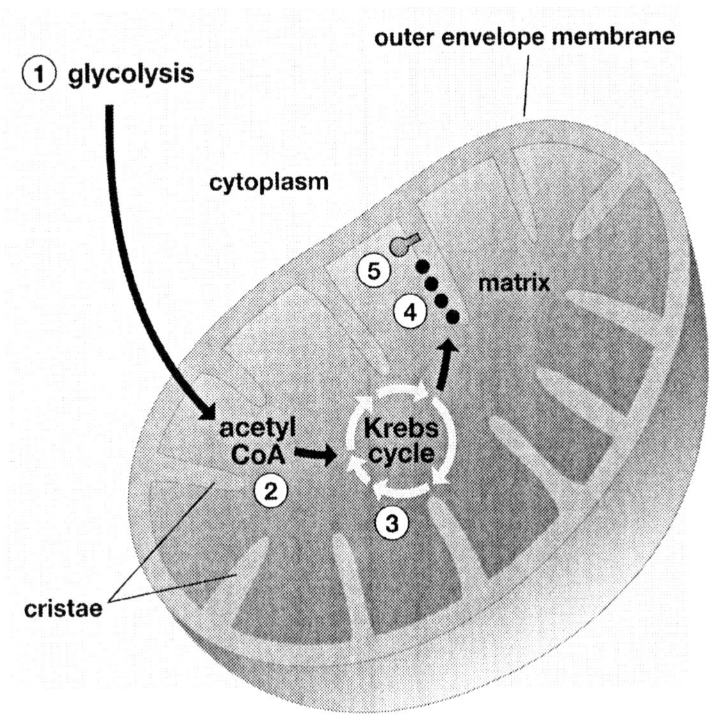

Figure 5.22 **Locations of the five respiration processes**

NOTES

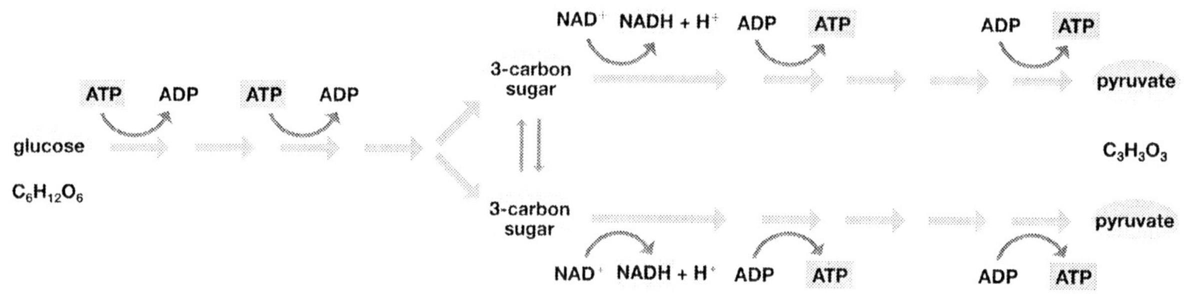

Figure 5.23 **Glycosis**

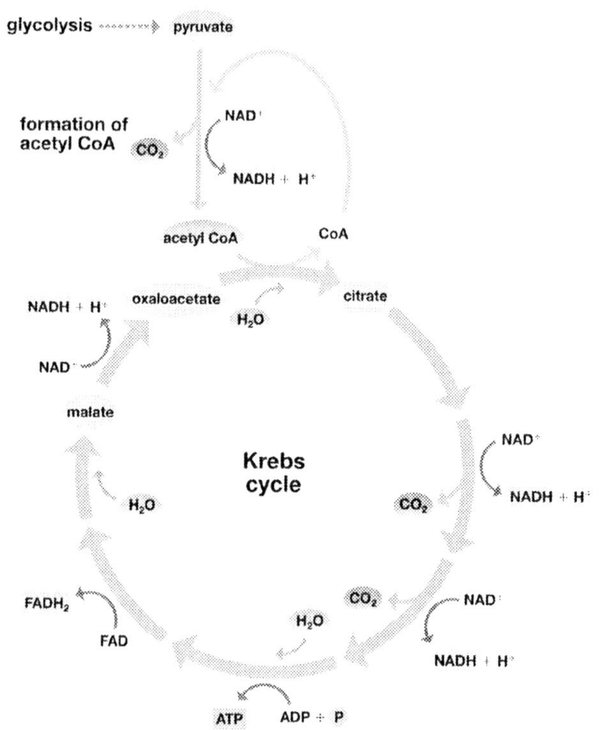

Figure 5.25 **Acetyl CoA formation and the Krebs cycle**

NOTES

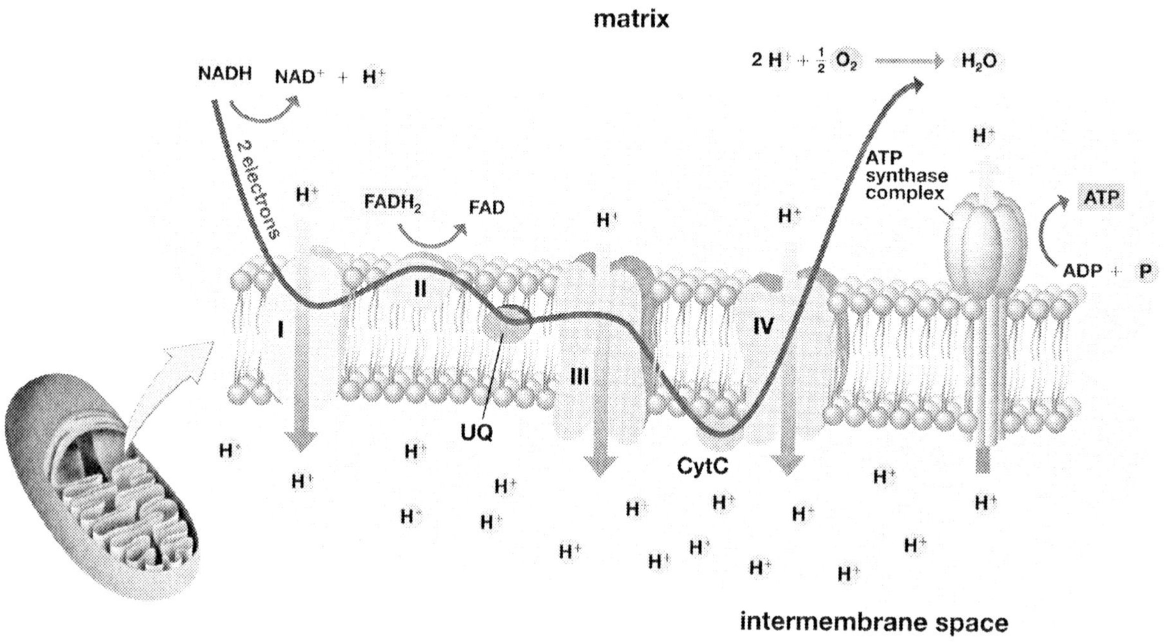

Figure 5.26 **The electron transport chain, chemiosmosis, and ATP synthase**

NOTES

CHAPTER 6

DNA and RNA

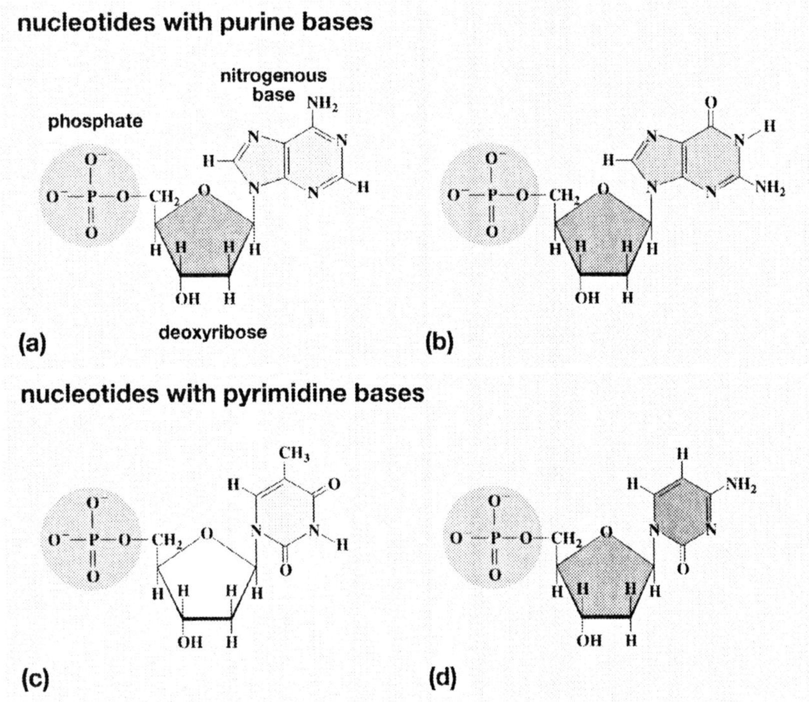

Figure 6.1 **The four nucleotides found in DNA**

NOTES

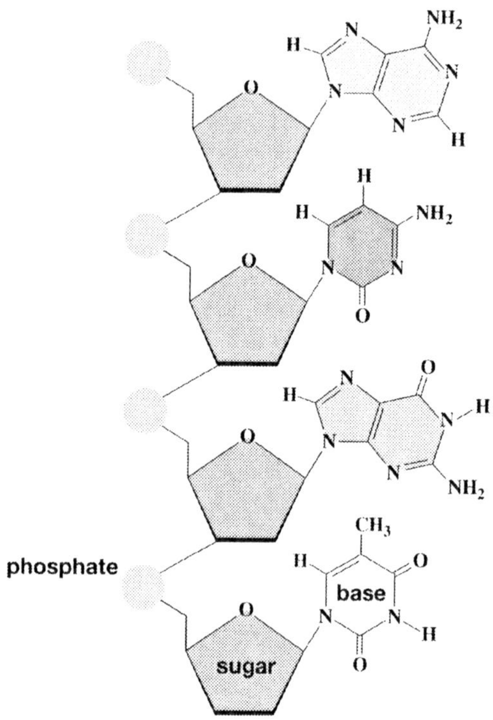

Figure 6.2 **A single nucleotide chain (of DNA)**

NOTES

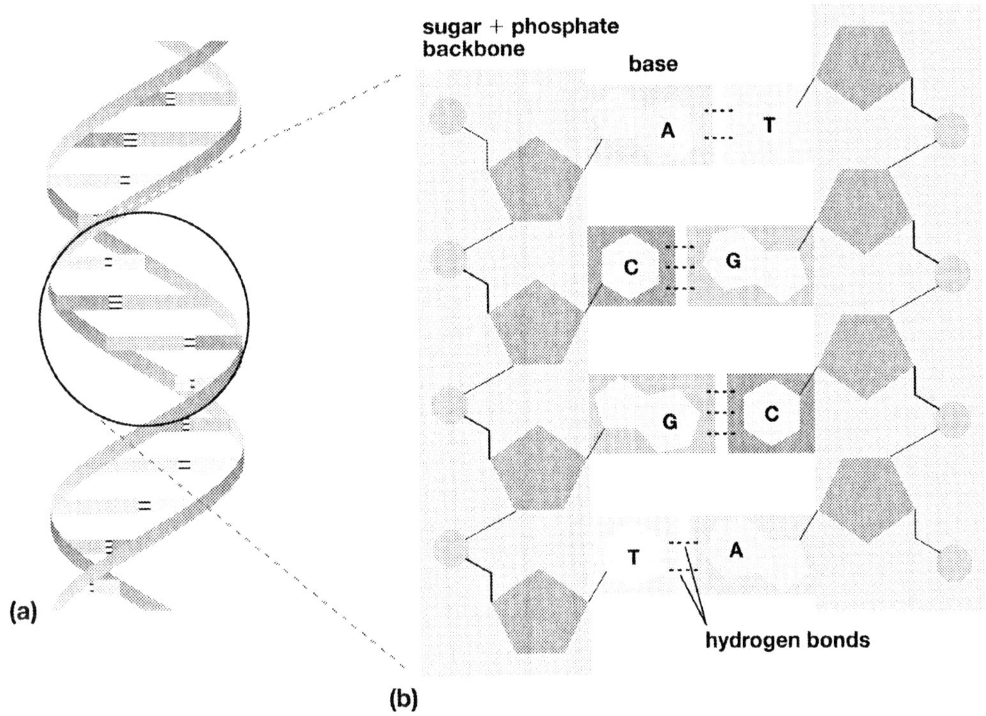

Figure 6.3 **Structure of DNA**

NOTES

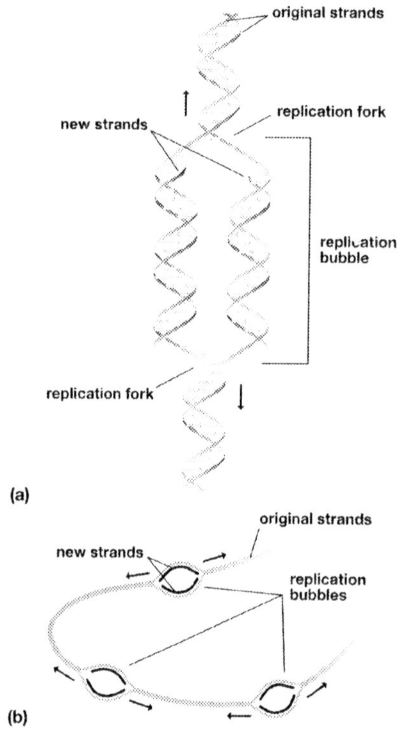

Figure 6.4 **Replication of DNA**

NOTES

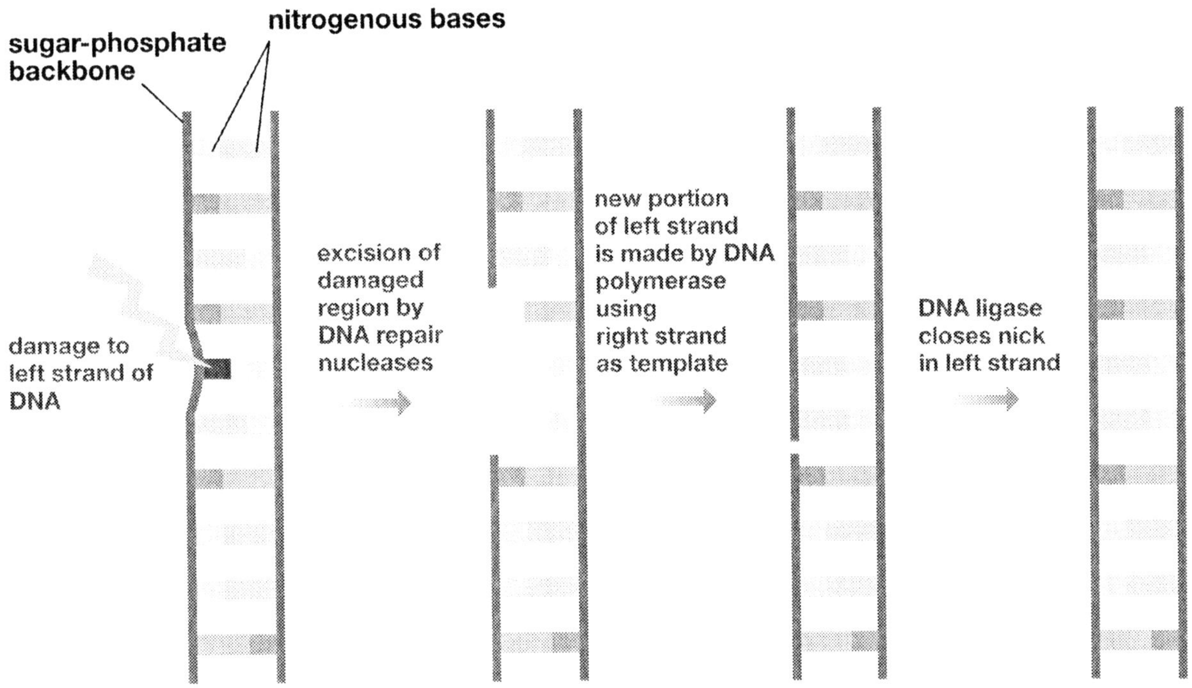

Figure 6.5 **Generalized mechanism of DNA repair**

NOTES

First position in codon	Second position in codon U	C	A	G	Third position in codon
U	phenylalanine	serine	tyrosine	cysteine	U
	phenylalanine	serine	tyrosine	cysteine	C
	leucine	serine	stop (UAA)	stop (UGA)	A
	leucine	serine	stop (UAG)	tryptophan	G
C	leucine	proline	histidine	arginine	U
	leucine	proline	histidine	arginine	C
	leucine	proline	glutamine	arginine	A
	leucine	proline	glutamine	arginine	G
A	isoleucine	threonine	asparagine	serine	U
	isoleucine	threonine	asparagine	serine	C
	isoleucine	threonine	lysine	arginine	A
	methionine (AUG)	threonine	lysine	arginine	G
G	valine	alanine	aspartic acid	glycine	U
	valine	alanine	aspartic acid	glycine	C
	valine	alanine	glutamic acid	glycine	A
	valine	alanine	glutamic acid	glycine	G

Figure 6.6 **The genetic code**

NOTES

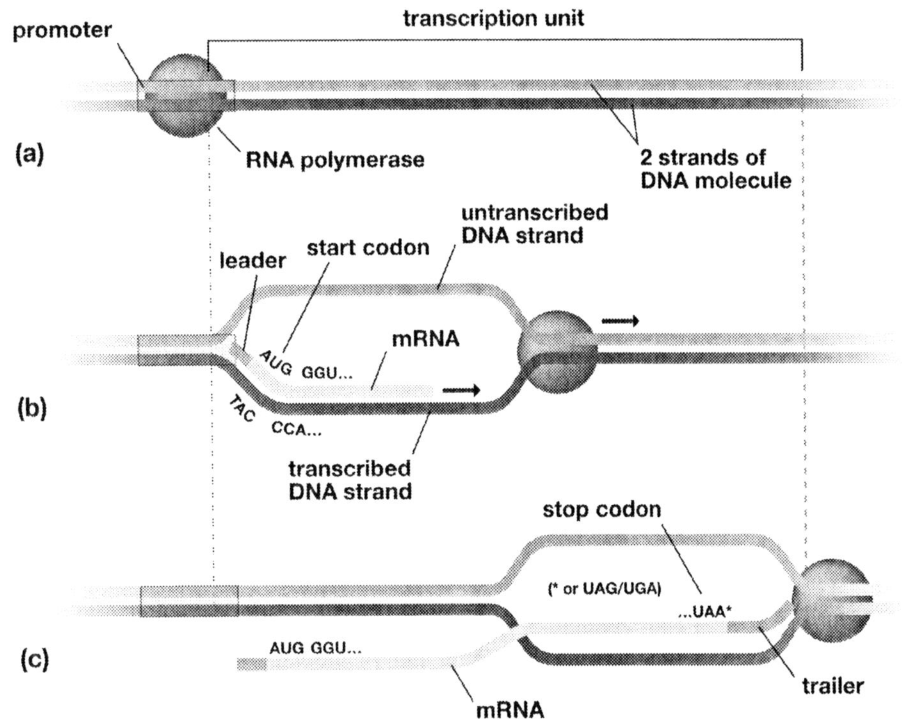

Figure 6.7 **Transcription of mRNA from a template strand of DNA**

NOTES

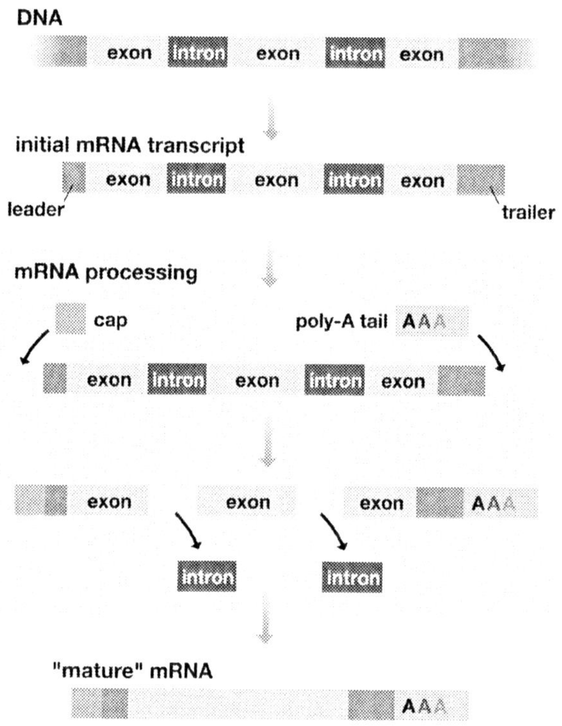

Figure 6.8 **The processing of a primary mRNA transcript in the eukaryotic nucleus**

NOTES

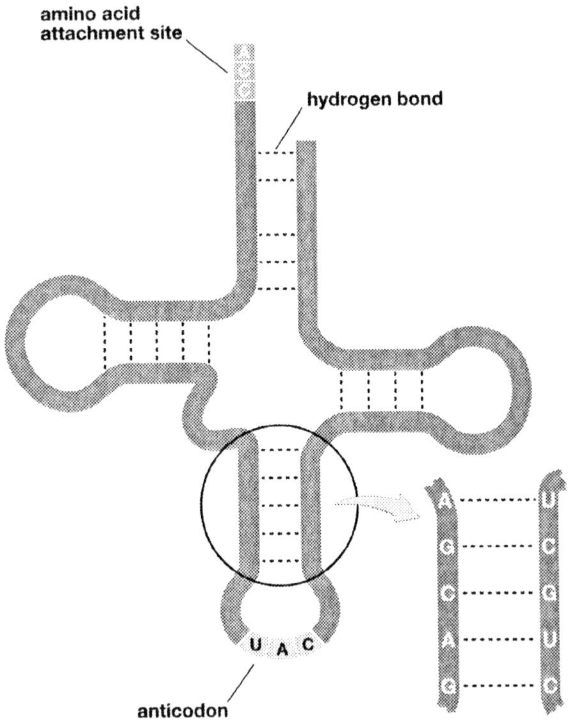

Figure 6.9 **Diagram of the structure of a tRNA molecule**

NOTES

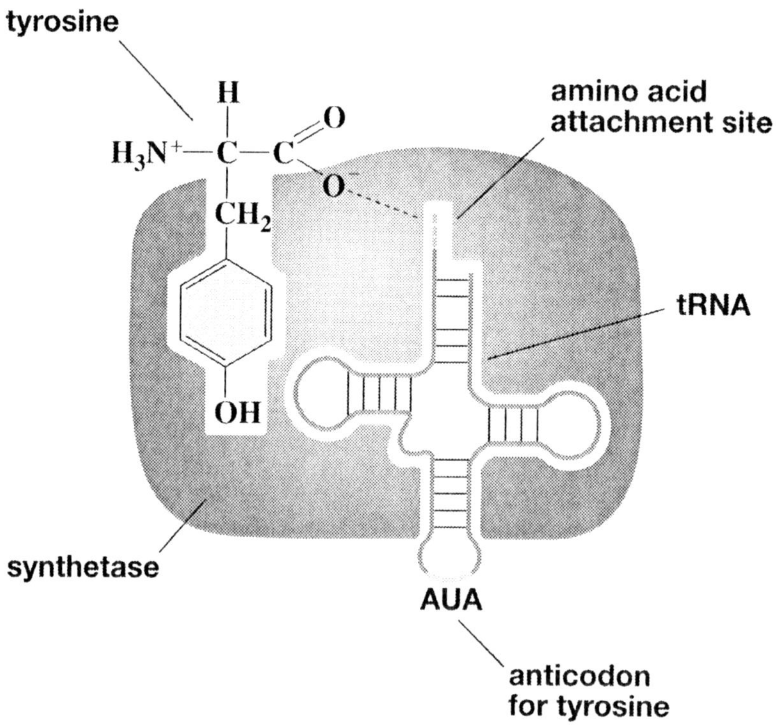

Figure 6.10 **tRNA molecules are linked to their appropriate amino acids by enzymes called aminoacyl-tRNA synthetases**

NOTES

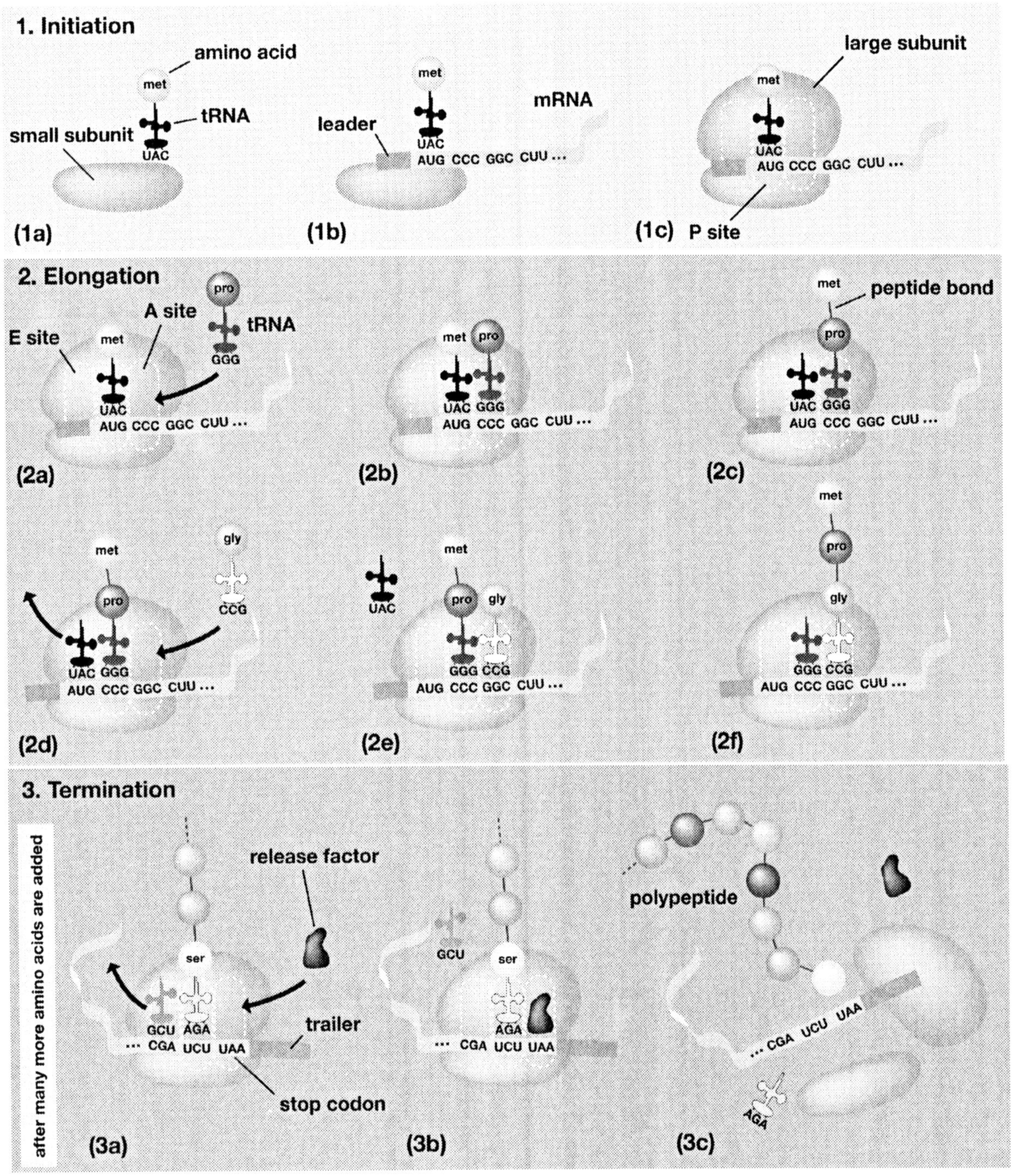

Figure 6.11 **Translation of an mRNA into a polypeptide chain in a protein**

NOTES

NOTES

CHAPTER 7

Cell Division

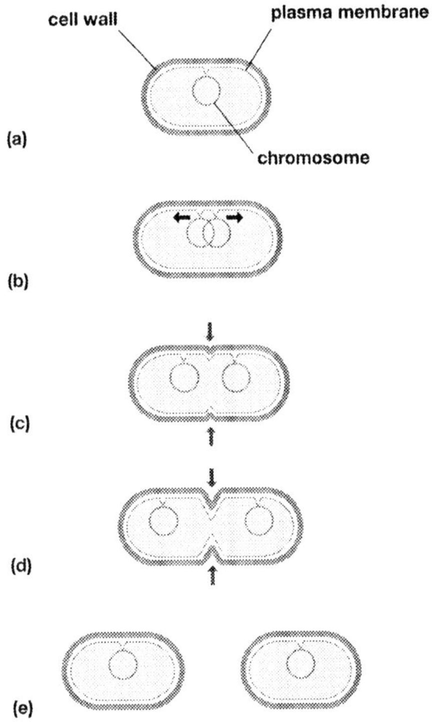

Figure 7.5 **Binary fission in bacteria**

NOTES

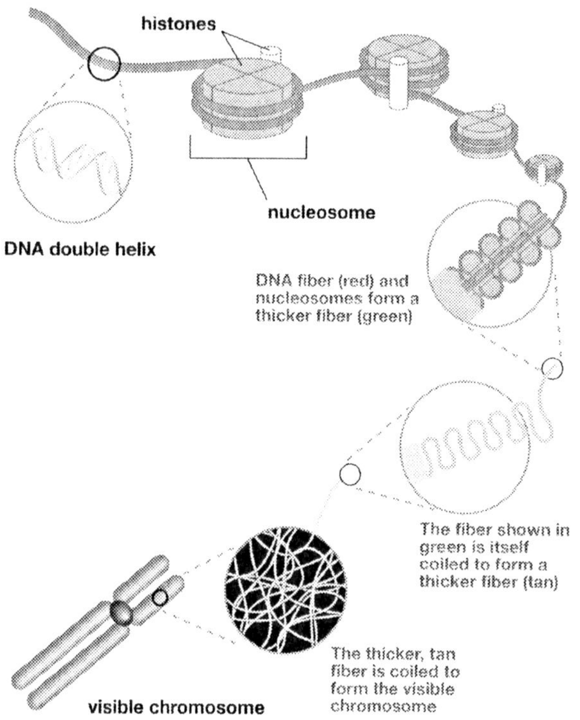

Figure 7.7 **Nucleosome and chromosome structure**

NOTES

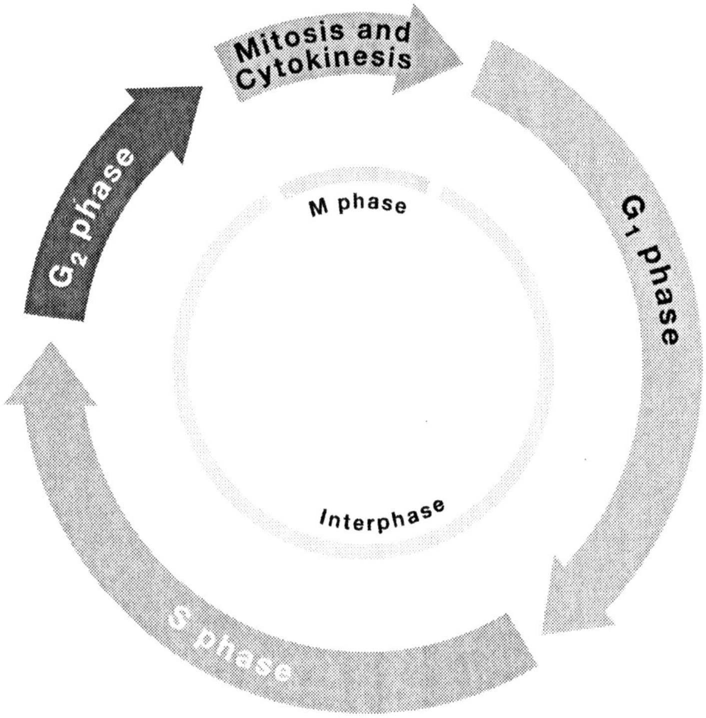

Figure 7.8 **The cell cycle**

NOTES

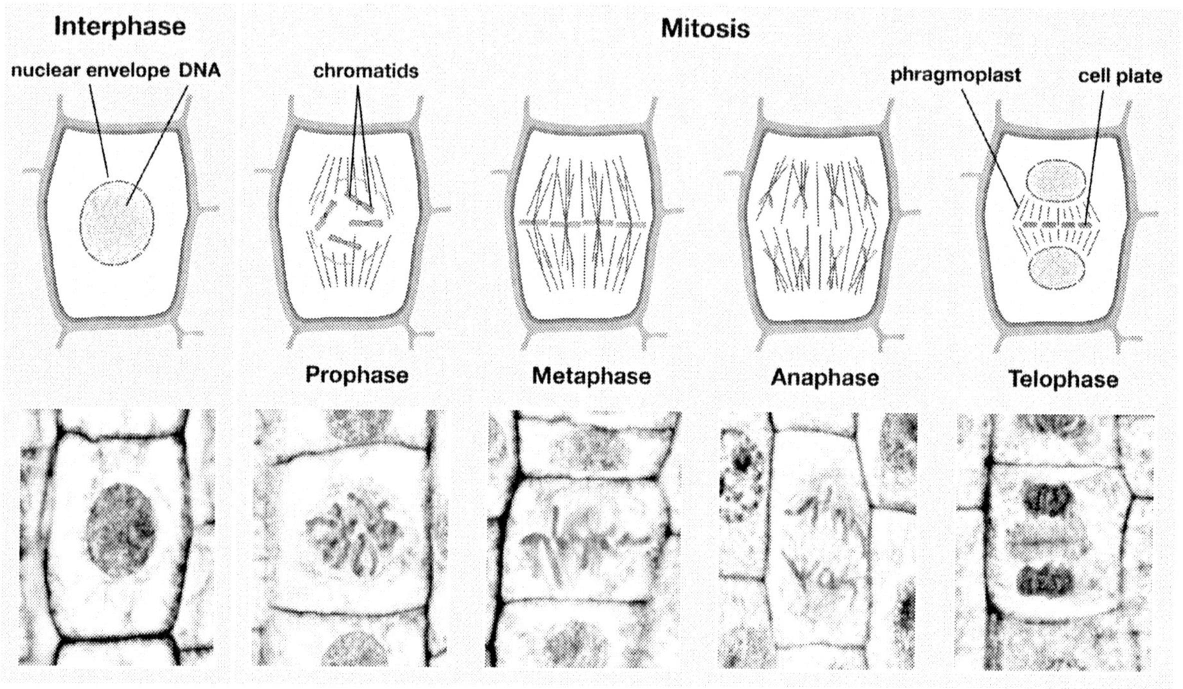

Figure 7.10 **The phases of mitosis**

NOTES

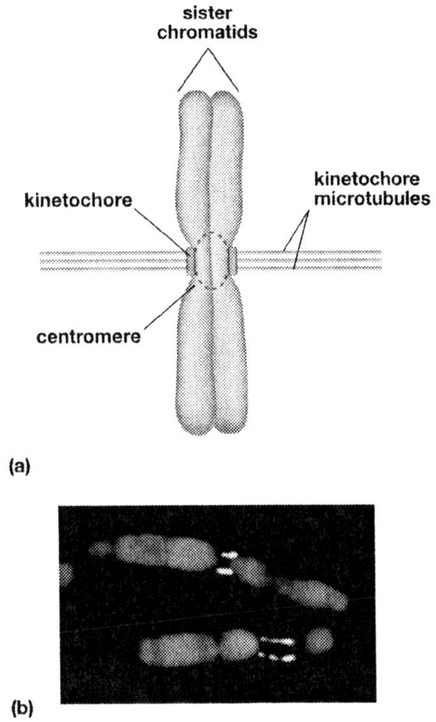

Figure 7.12 **Chromosome and kinetochores**

NOTES

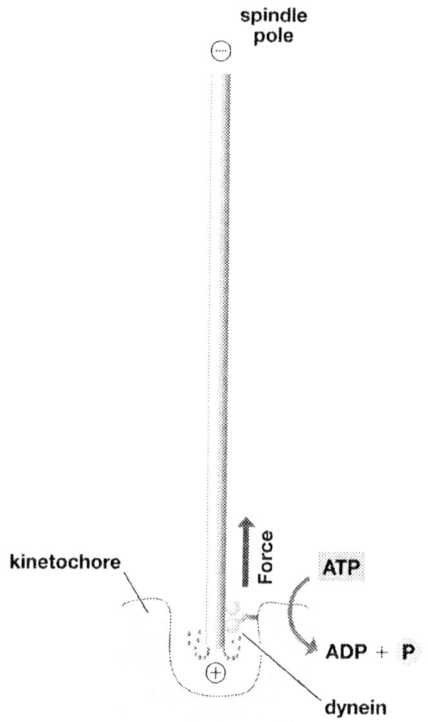

Figure 7.14 **The motor protein model of how chromosomes might move to the poles**

NOTES

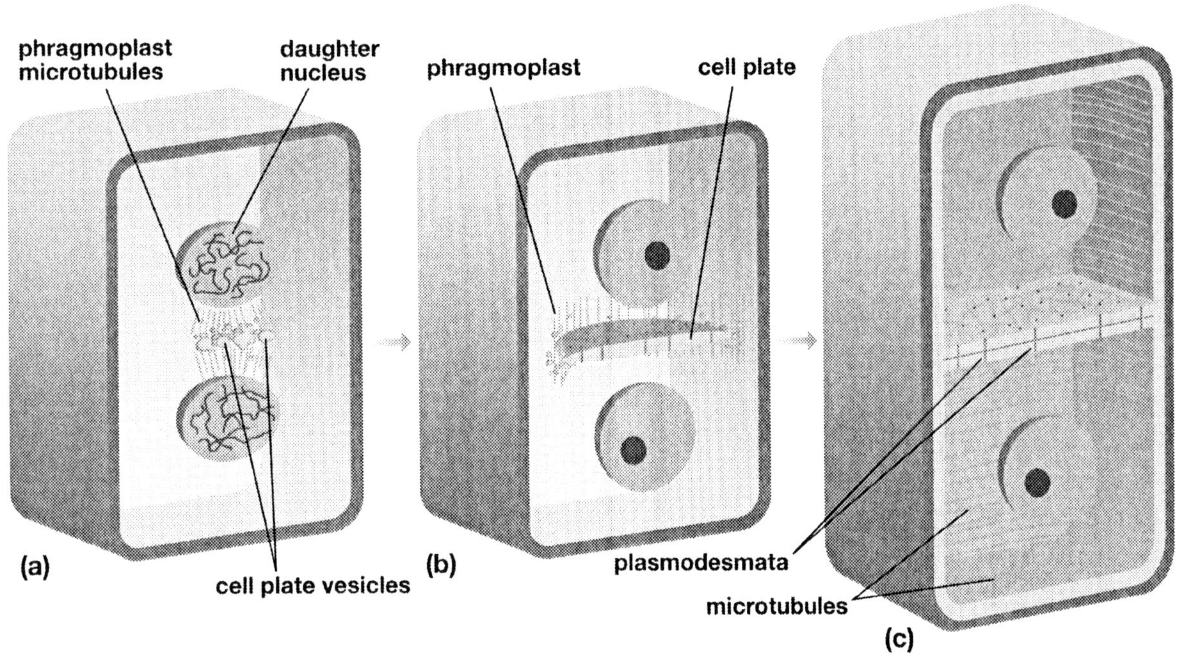

Figure 7.15 **Diagram of plant cytokinesis phases**

NOTES

NOTES

CHAPTER 8

Plant Structure, Growth, and Development

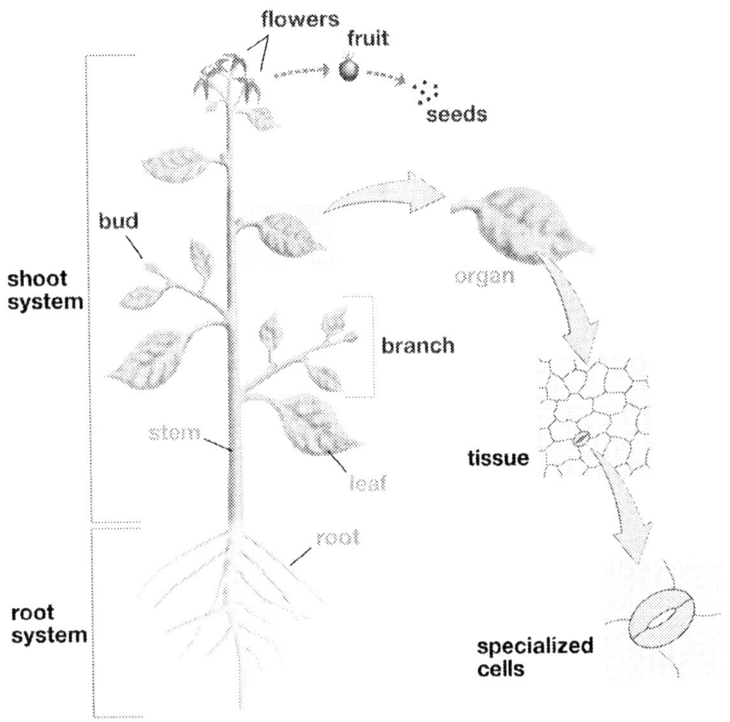

Figure 8.2 **The plant body and its component organs, tissues, and specialized cells**

NOTES

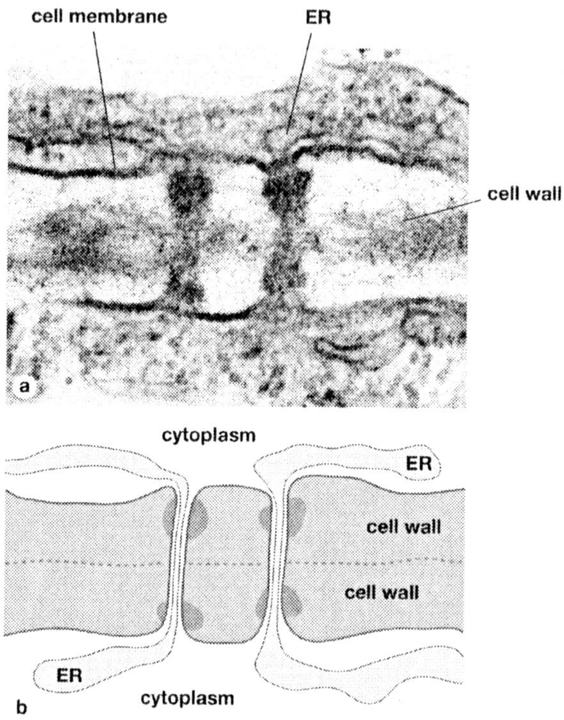

Figure 8.4 **Plant plasmodesmata**

NOTES

Figure 8.5 **Plant primary meristems**

NOTES

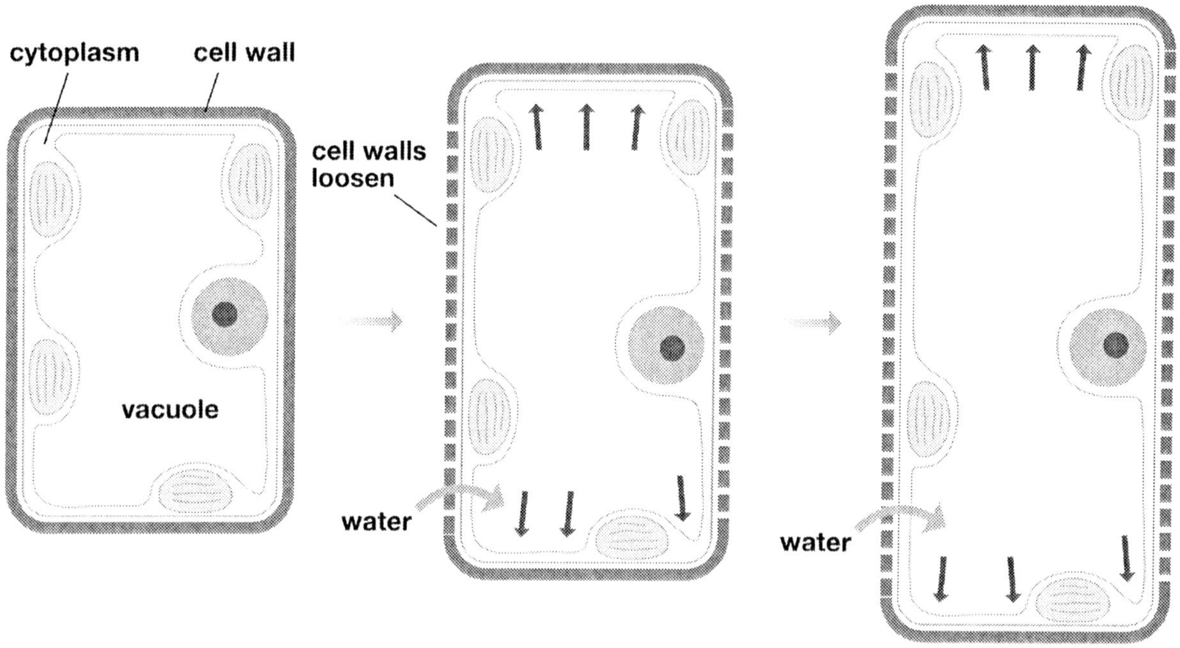

Figure 8.9 **Growth of plant cells through water uptake**

NOTES

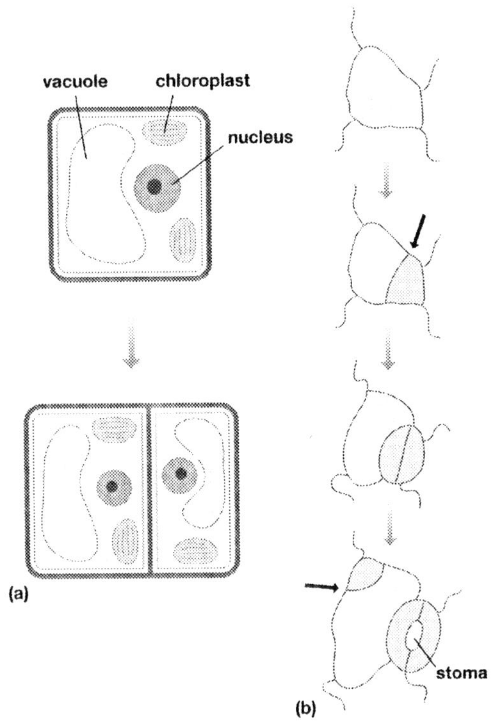

Figure 8.16 **Plant cell differentiation by unequal cell division**

NOTES

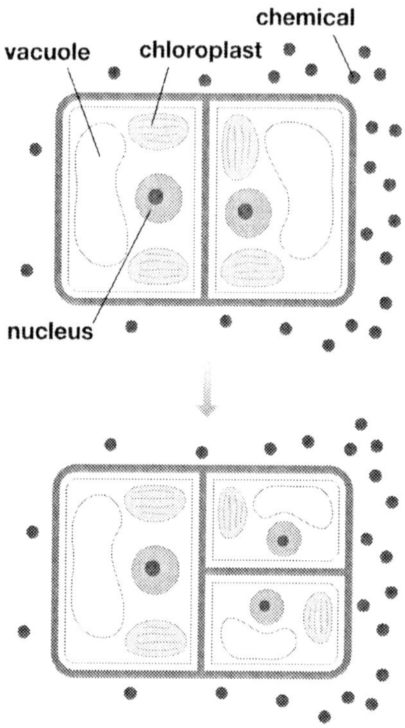

Figure 8.17 **Plant cell differentiation may result from position effects**

NOTES

CHAPTER 9

Stems and Materials Transport

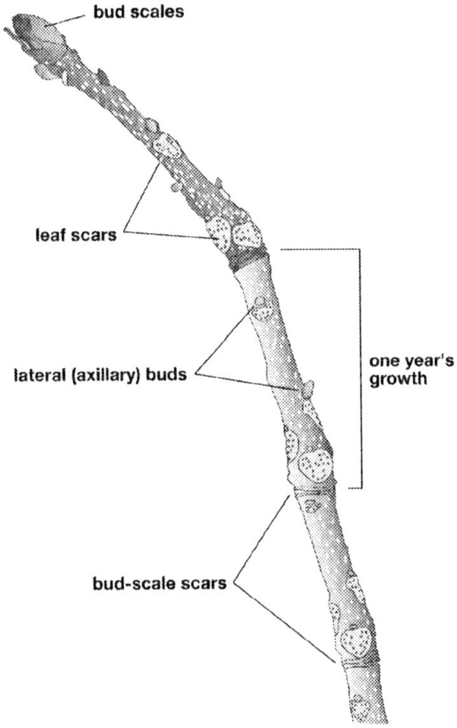

Figure 9.4 **Twig surfaces possess scars where leaves and bud scales were formerly attached**

NOTES

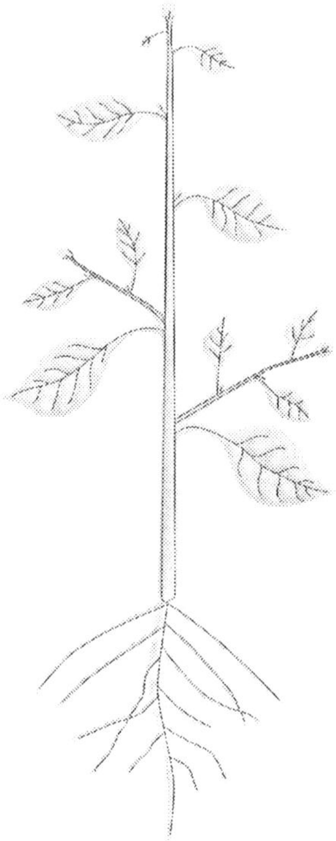

Figure 9.9 **Vascular system of a plant**

NOTES

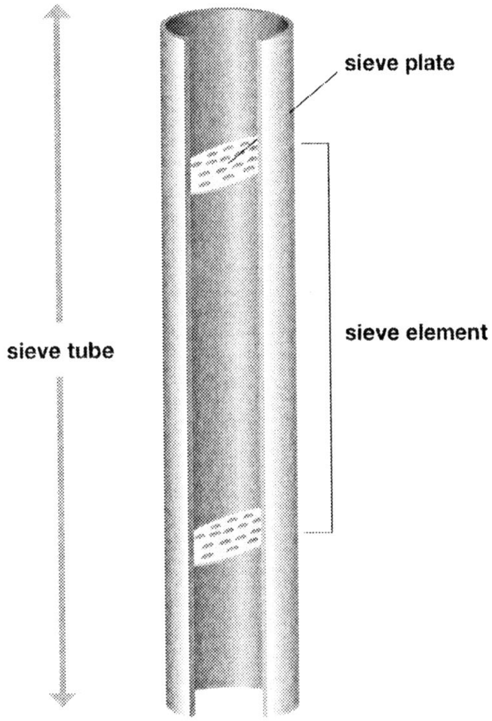

Figure 9.10 **Sieve elements forming sieve tube**

NOTES

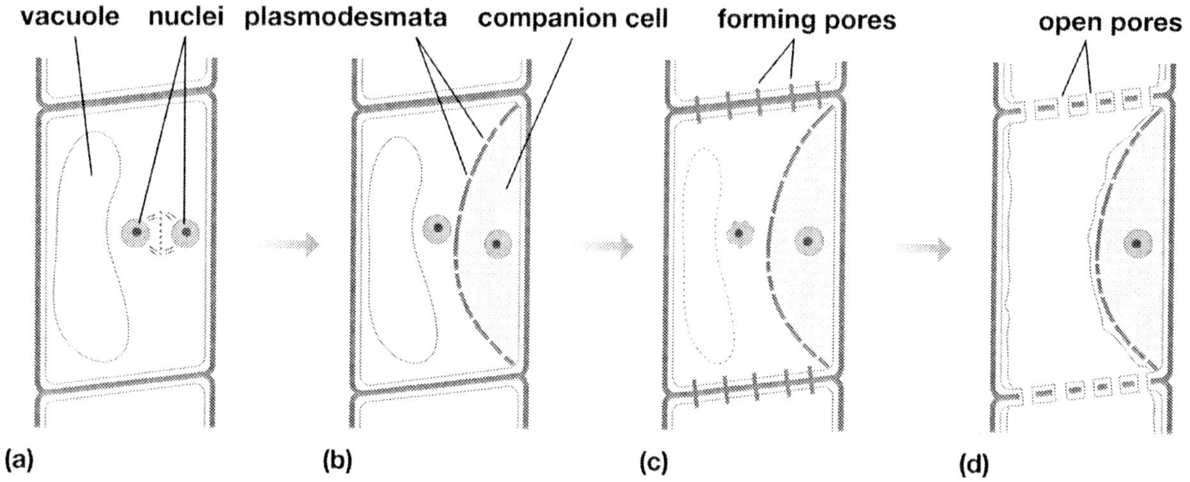

Figure 9.12 **Development of sieve element and companion cell after unequal cell division**

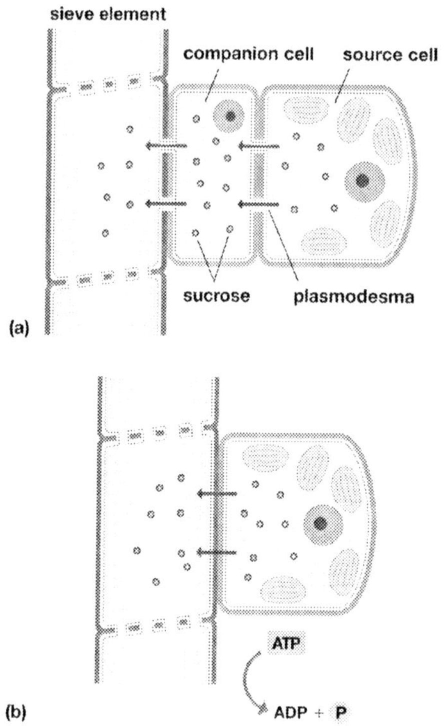

Figure 9.13 **Symplastic and apoplastic phloem loading**

NOTES

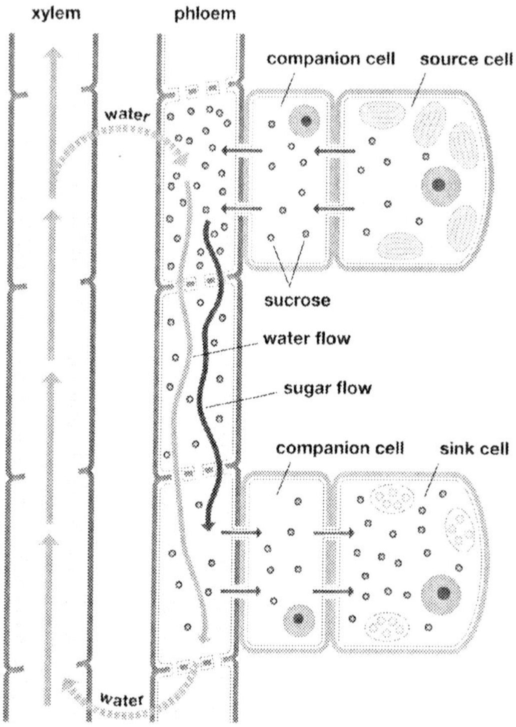

Figure 9.14 **Mass flow in phloem**

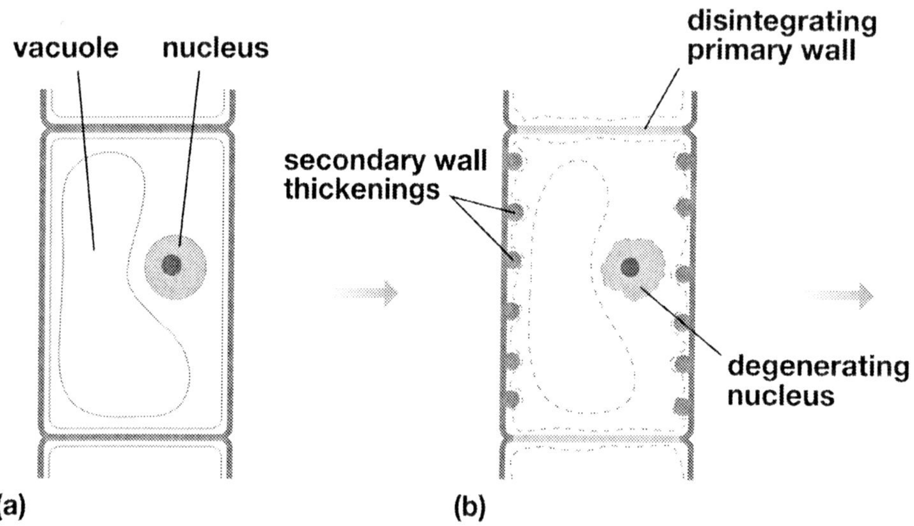

Figure 9.15 **Vessel development**

NOTES

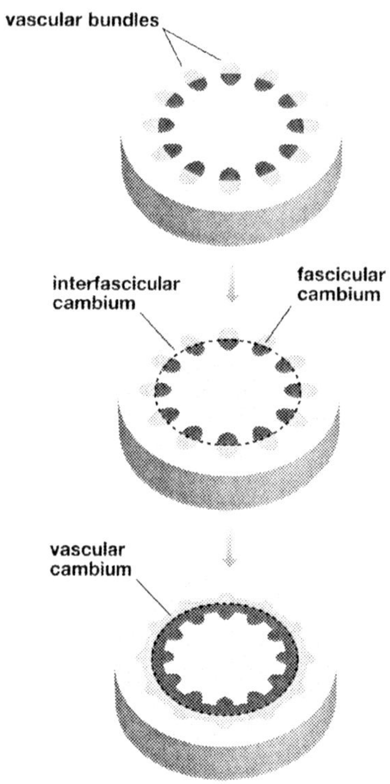

Figure 9.19 **Vascular cambium formation**

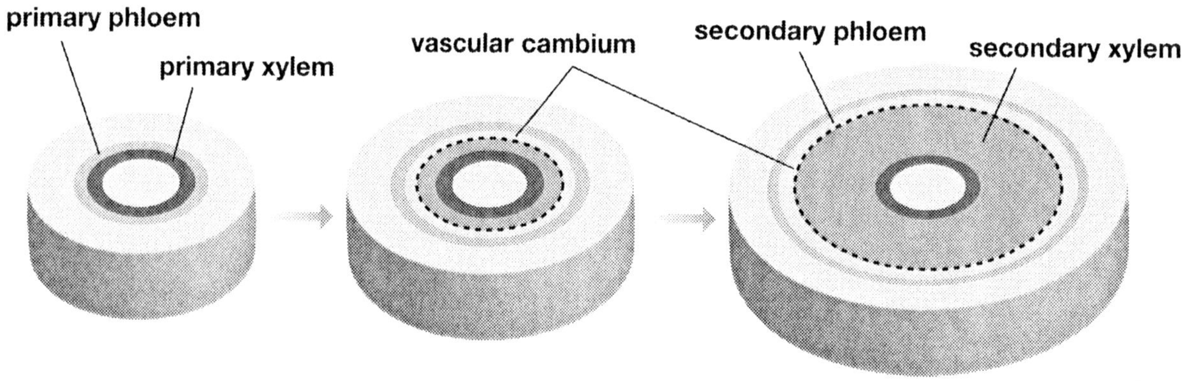

Figure 9.20 **Secondary xylem and phloem develop from vascular cambium**

NOTES

NOTES

CHAPTER 10

Roots and Plant Nutrition

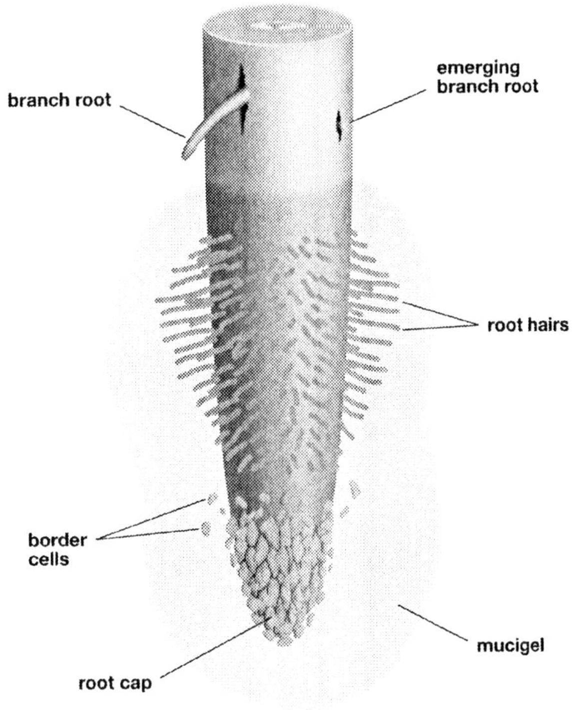

Figure 10.13 **External structure of a root**

NOTES

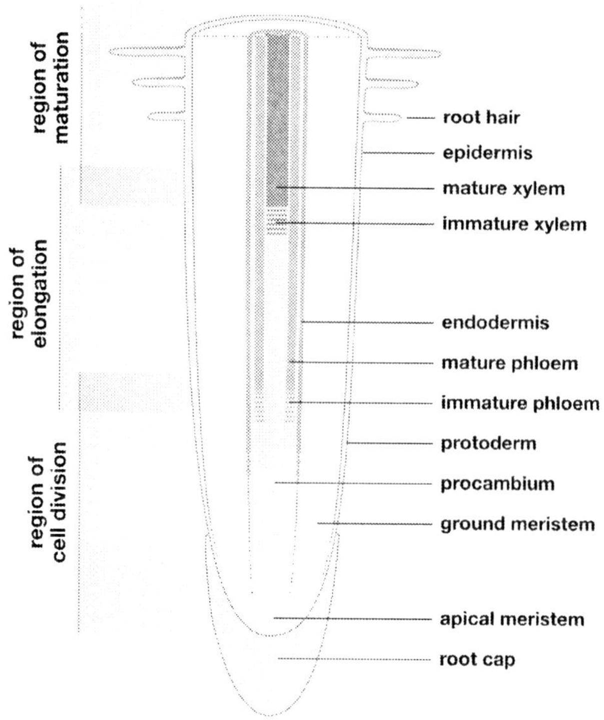

Figure 10.14 **Zones of root cap, meristematic cell division, cell elongation, and tissue maturation**

NOTES

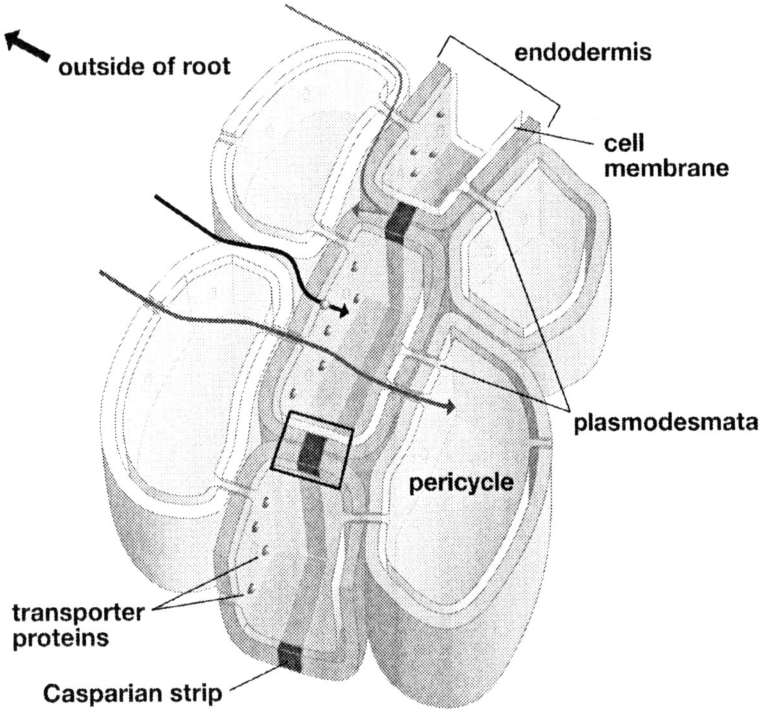

Figure 10.17 **Endodermis and Casparian strip**

NOTES

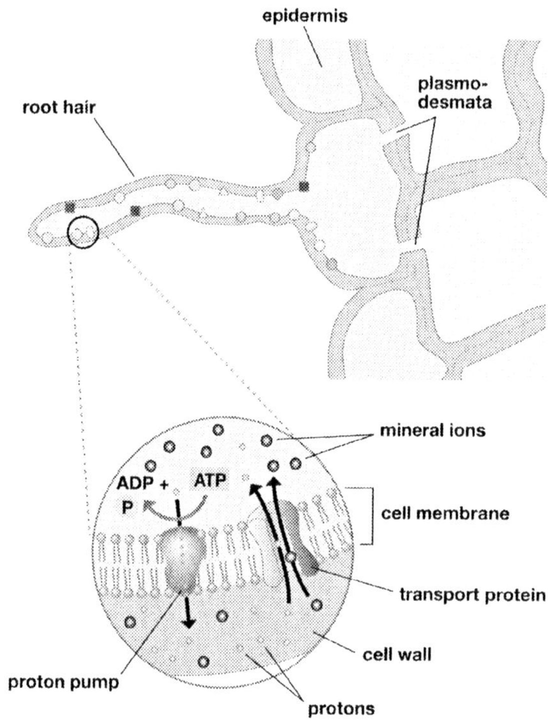

Figure 10.20 **Root hair with cell membrane transporters and plasmodesmata**

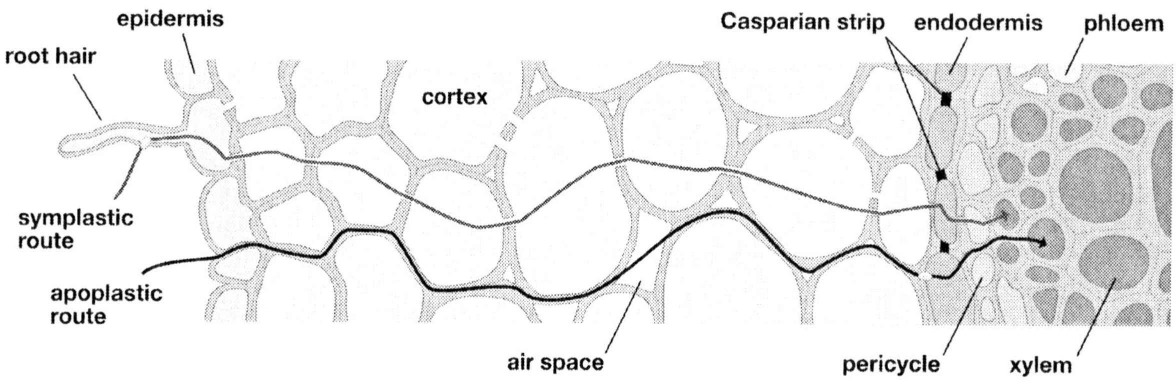

Figure 10.21 **Minerals can move from cell to cell through plasmodesmata or via cell walls and intercellular spaces en route from root hair to vascular tissue**

NOTES

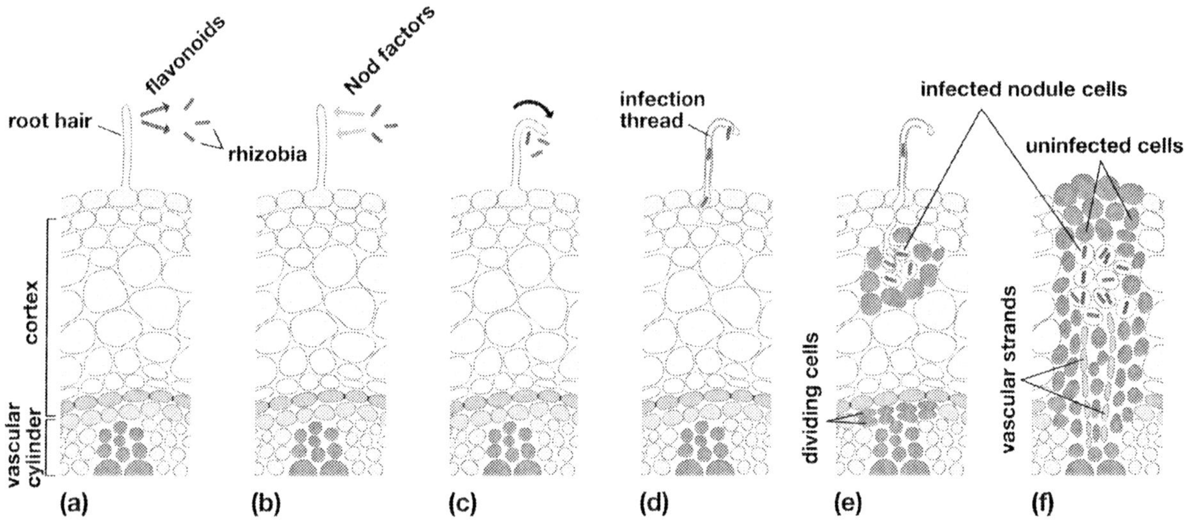

Figure 10.23 **Legume root nodule formation**

NOTES

CHAPTER 11

Leaves: Photosynthesis and Transportation

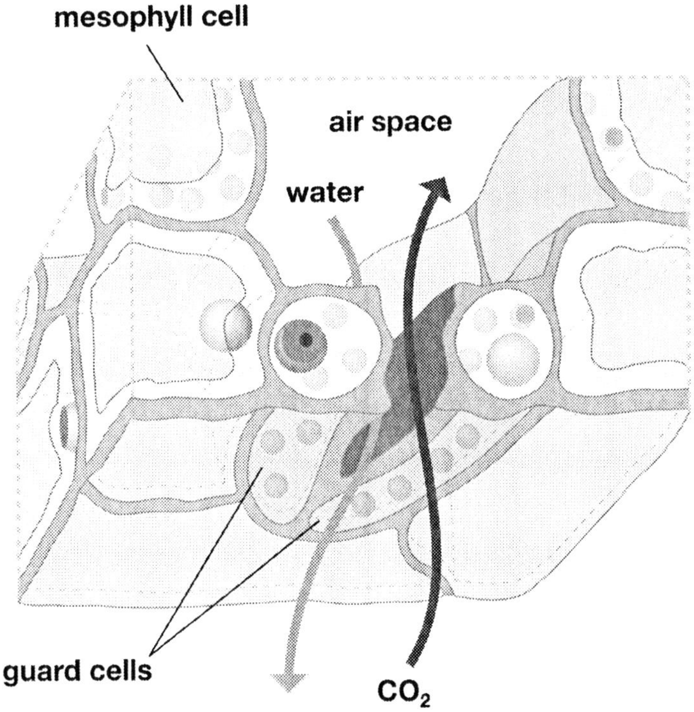

Figure 11.11 **Stoma**

NOTES

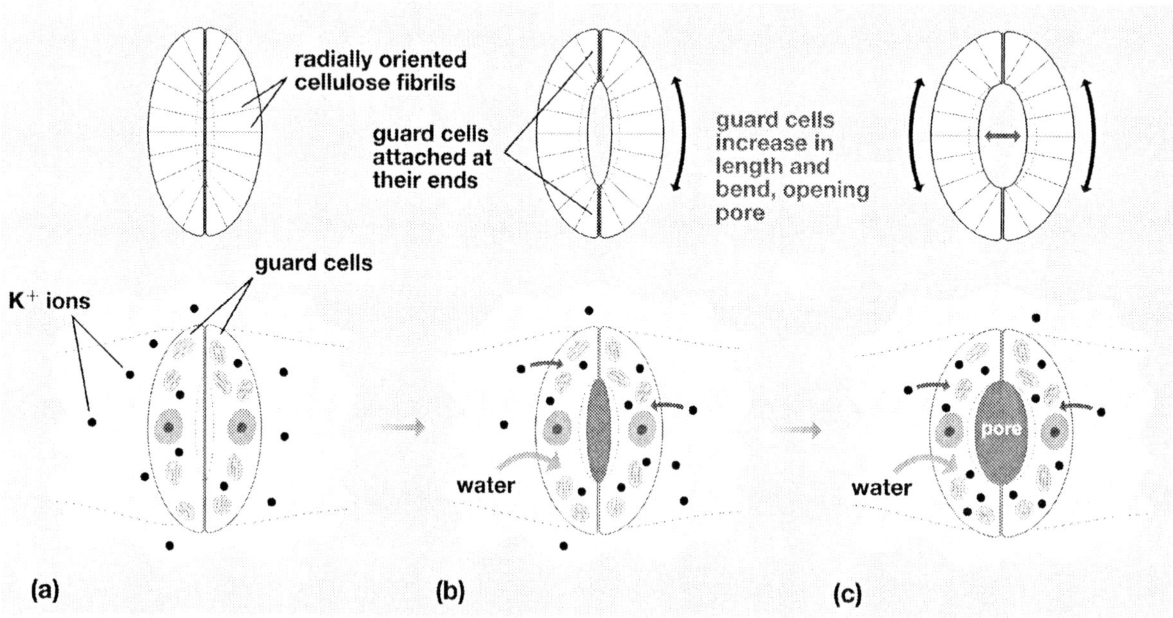

Figure 11.16 **Diagram of stomatal function**

NOTES

CHAPTER 12 Plant Behavior

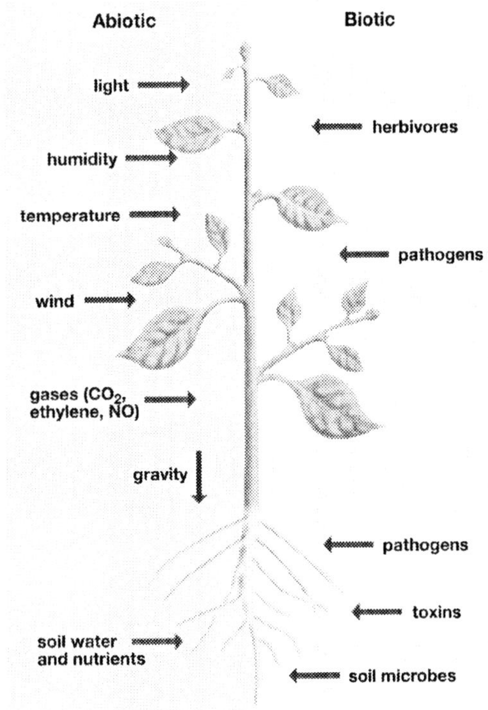

Figure 12.1 **Some of the abiotic signals received by plants**

NOTES

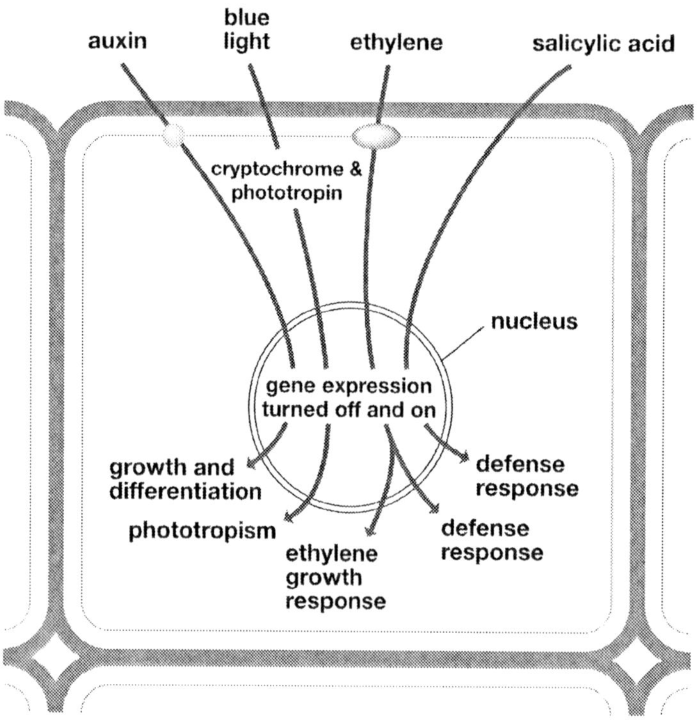

Figure 12.2 **Signal transduction**

NOTES

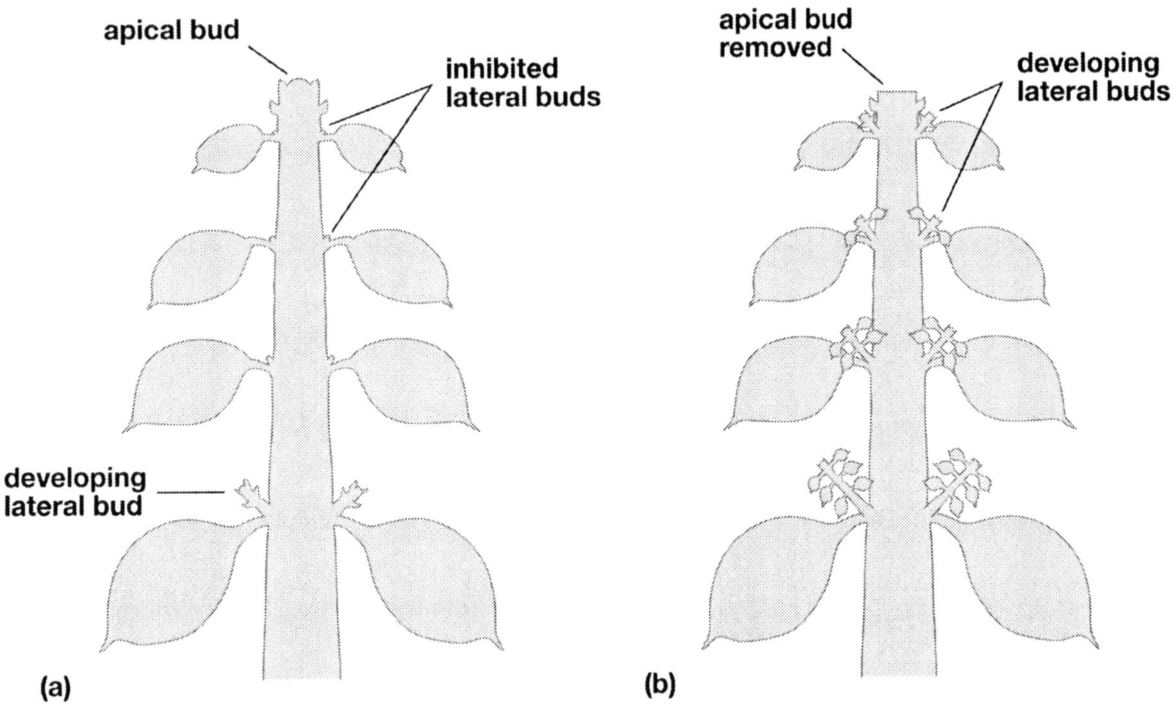

Figure 12.4 **Apical dominance**

NOTES

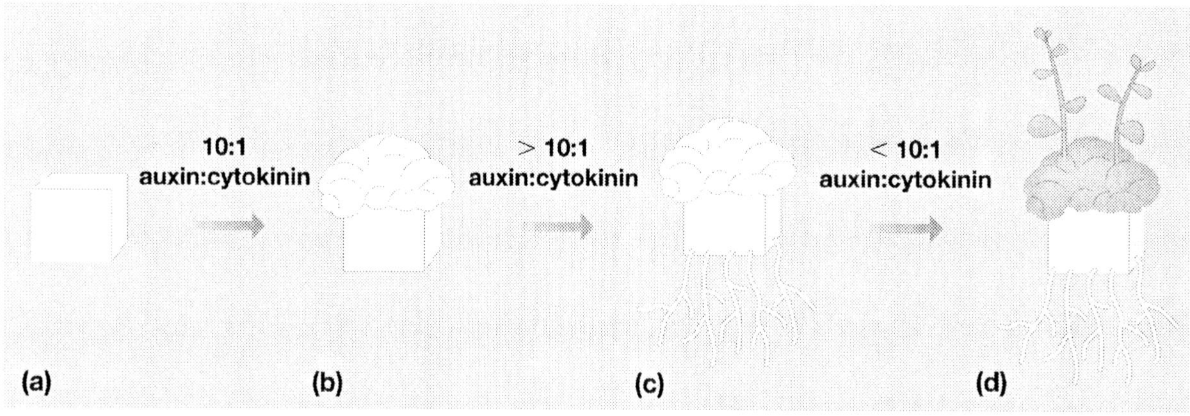

Figure 12.5 **Diagram showing the use of hormones in plant tissue culture**

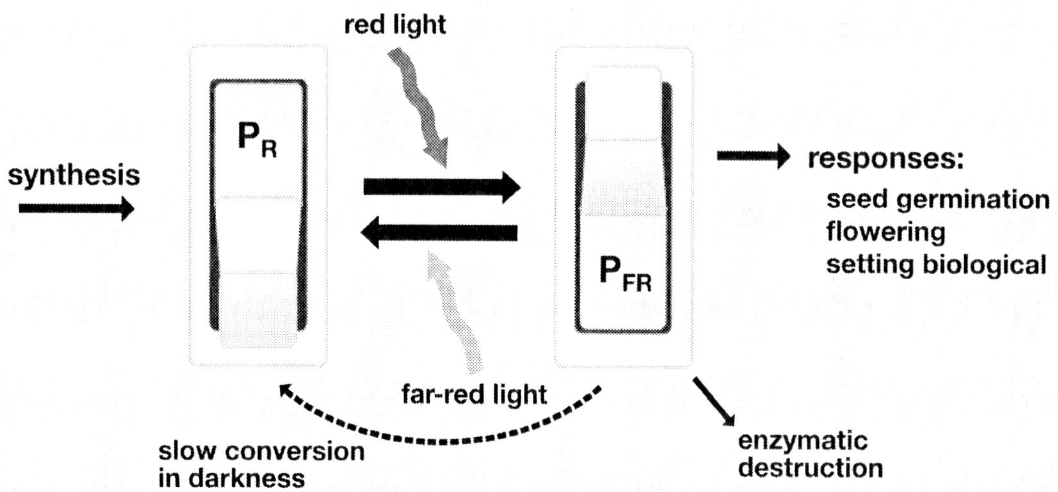

Figure 12.9 **Phytochrome enables plants to respond to red and far-red light**

NOTES

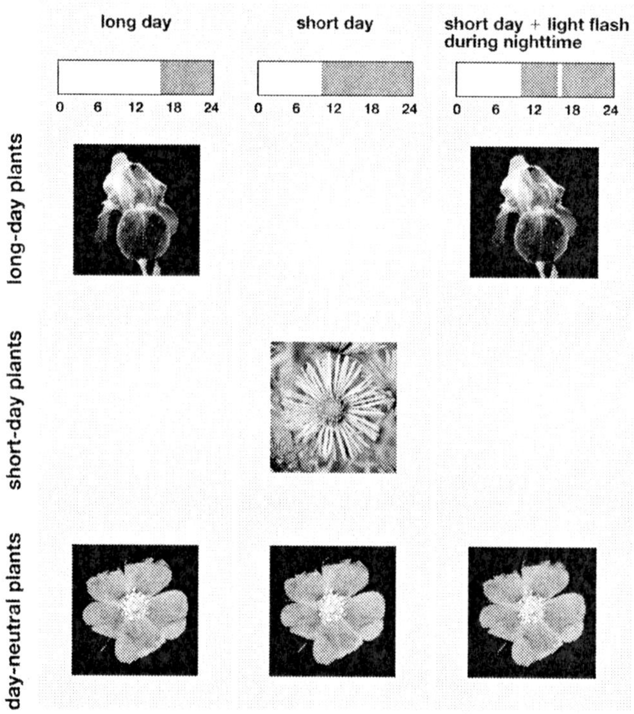

Figure 12.10 **Photoperiodism**

NOTES

CHAPTER 13

Reproduction, Meiosis, and Life Cycles

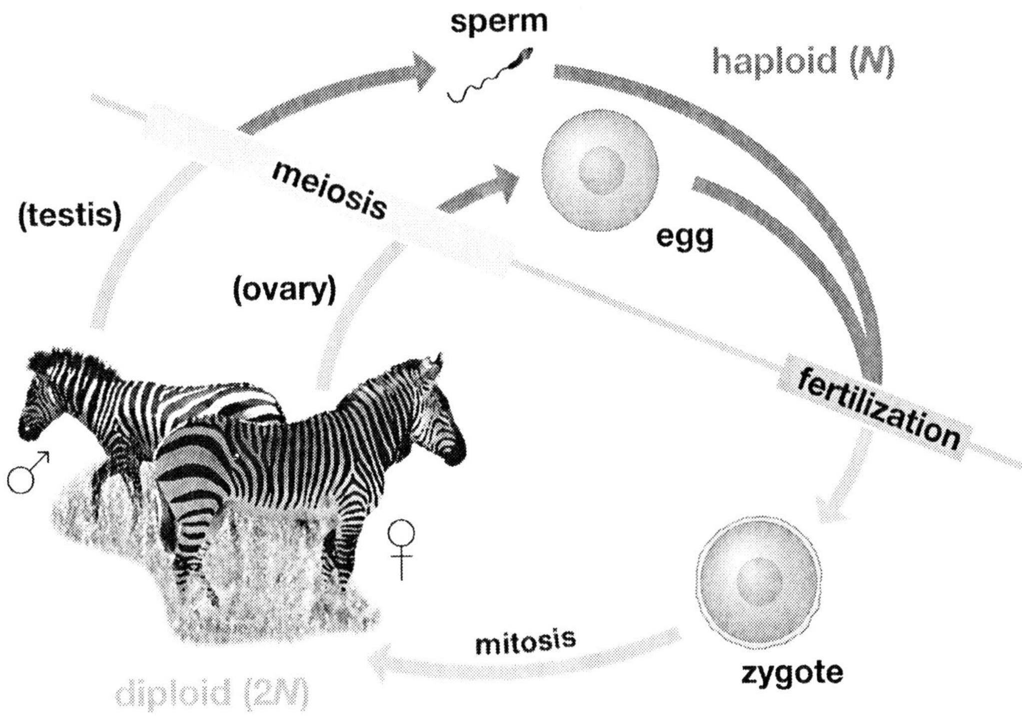

Figure 13.3 **The sexual reproductive cycle of animals**

NOTES

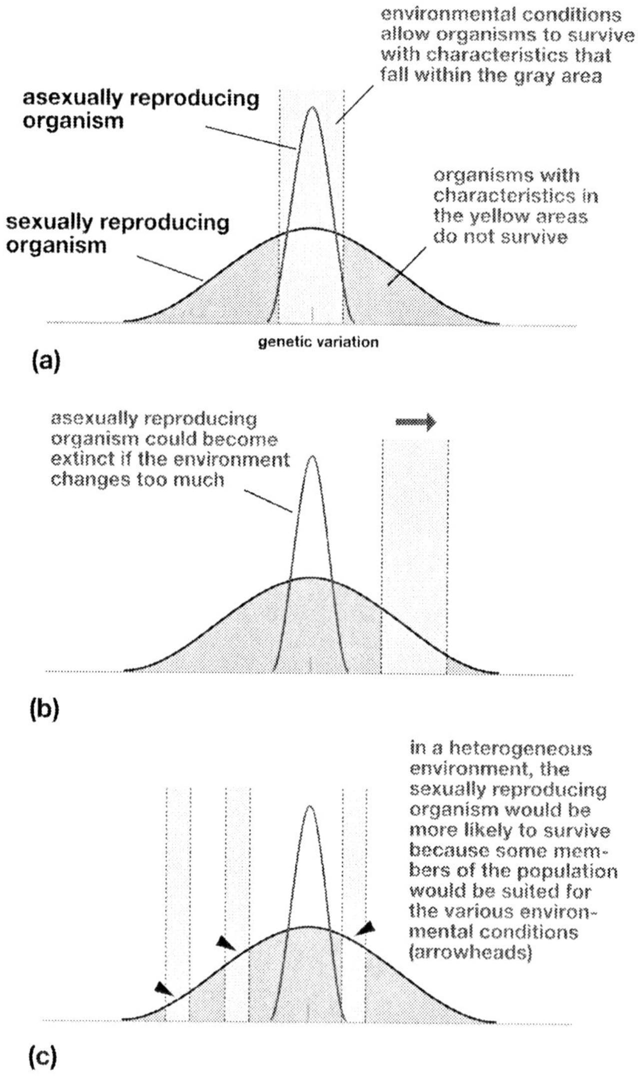

Figure 13.4 **Advantages of sexual and asexual reproduction**

NOTES

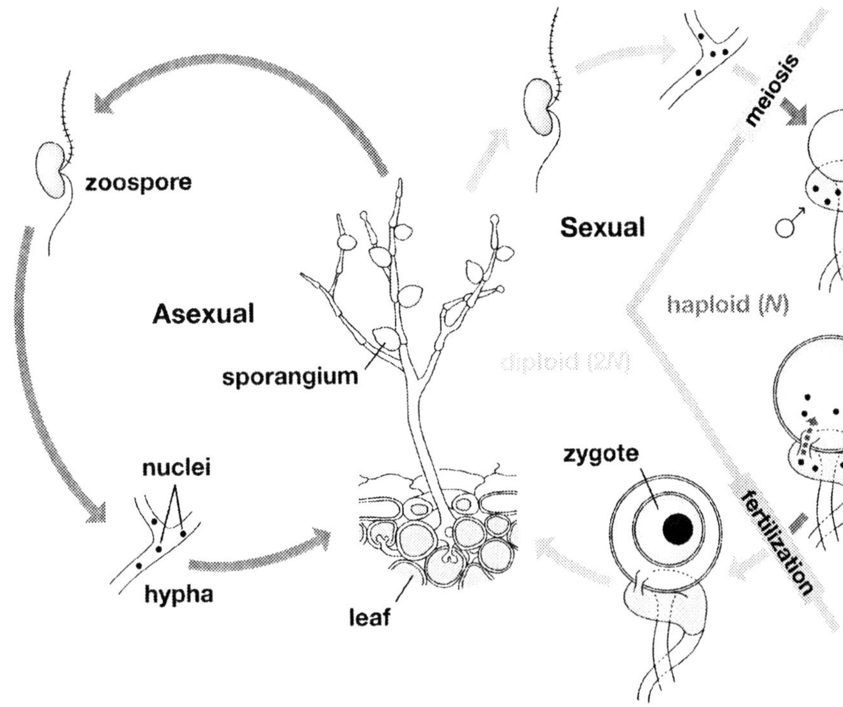

Figure 13.8 ***Phytophthora*, the plant destroyer**

NOTES

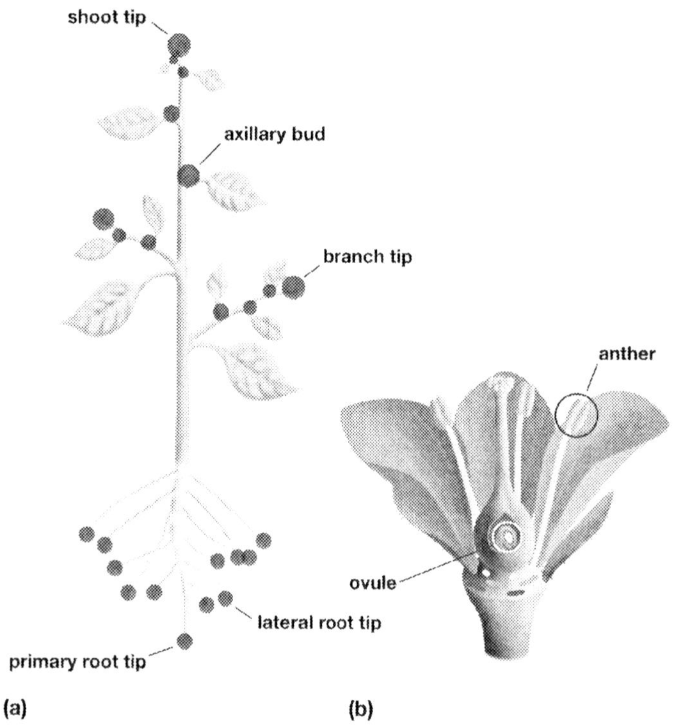

Figure 13.11 **Location of mitosis versus meiosis in plants**

NOTES

NOTES

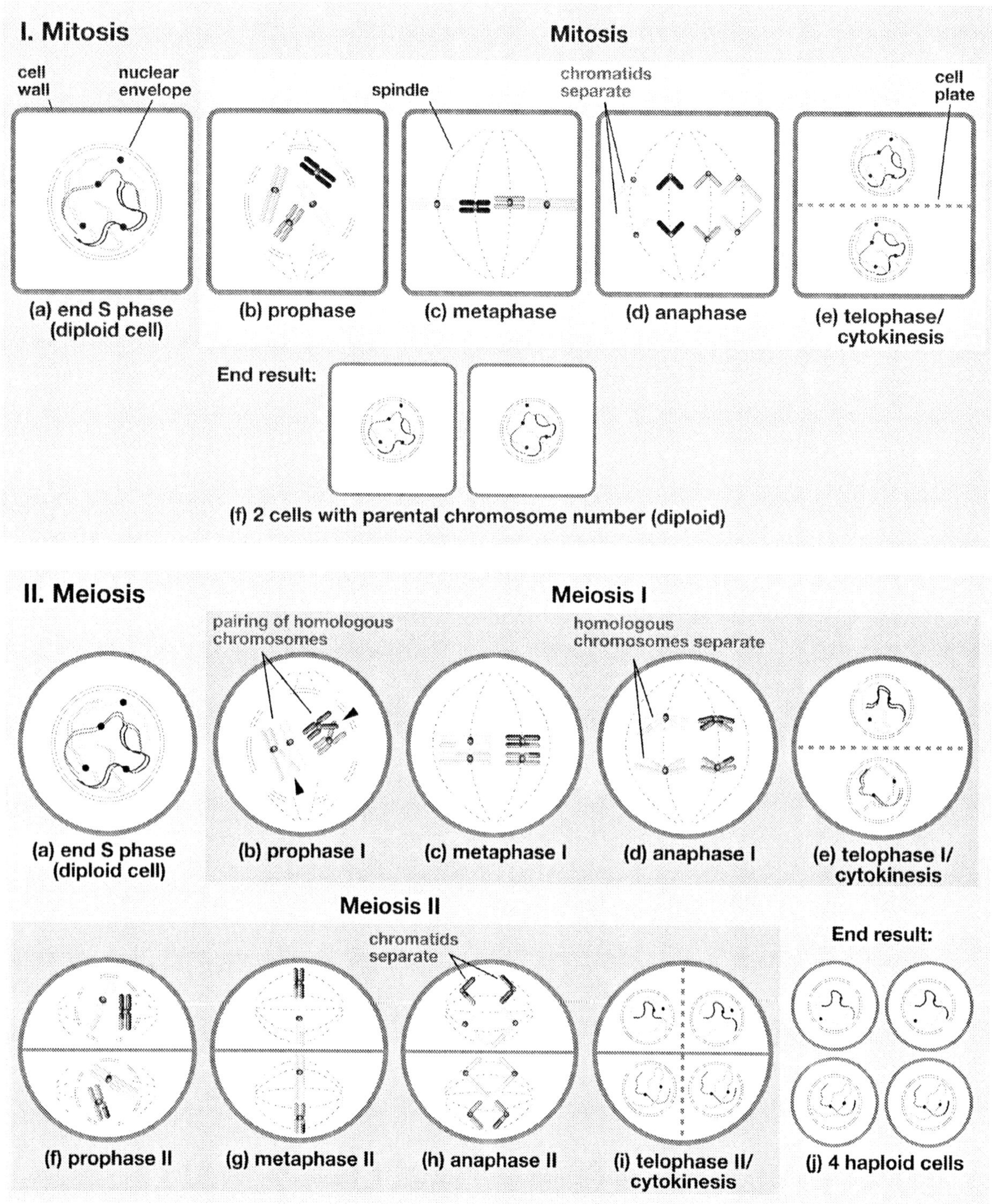

Figure 13.12 **Comparison of mitosis and meiosis**

NOTES

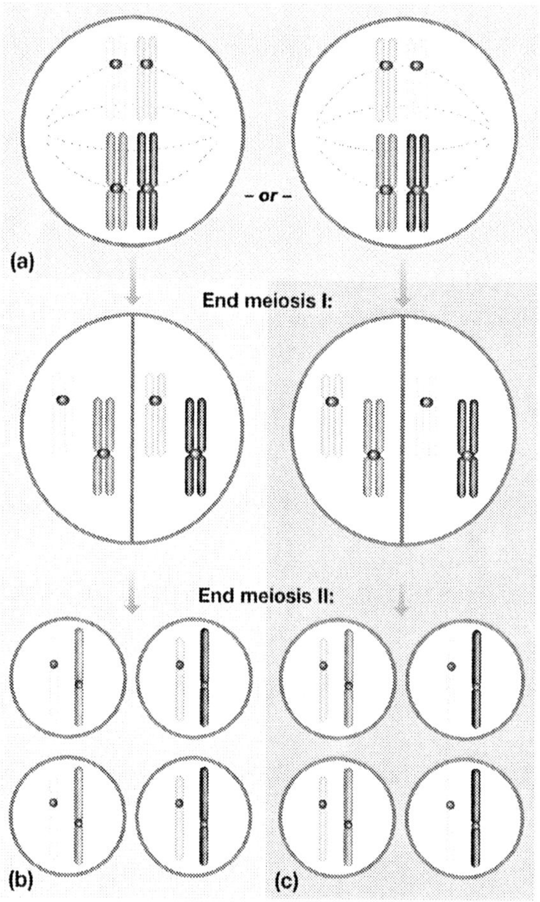

Figure 13.13 **Segregation of chromosomes in meiosis**

NOTES

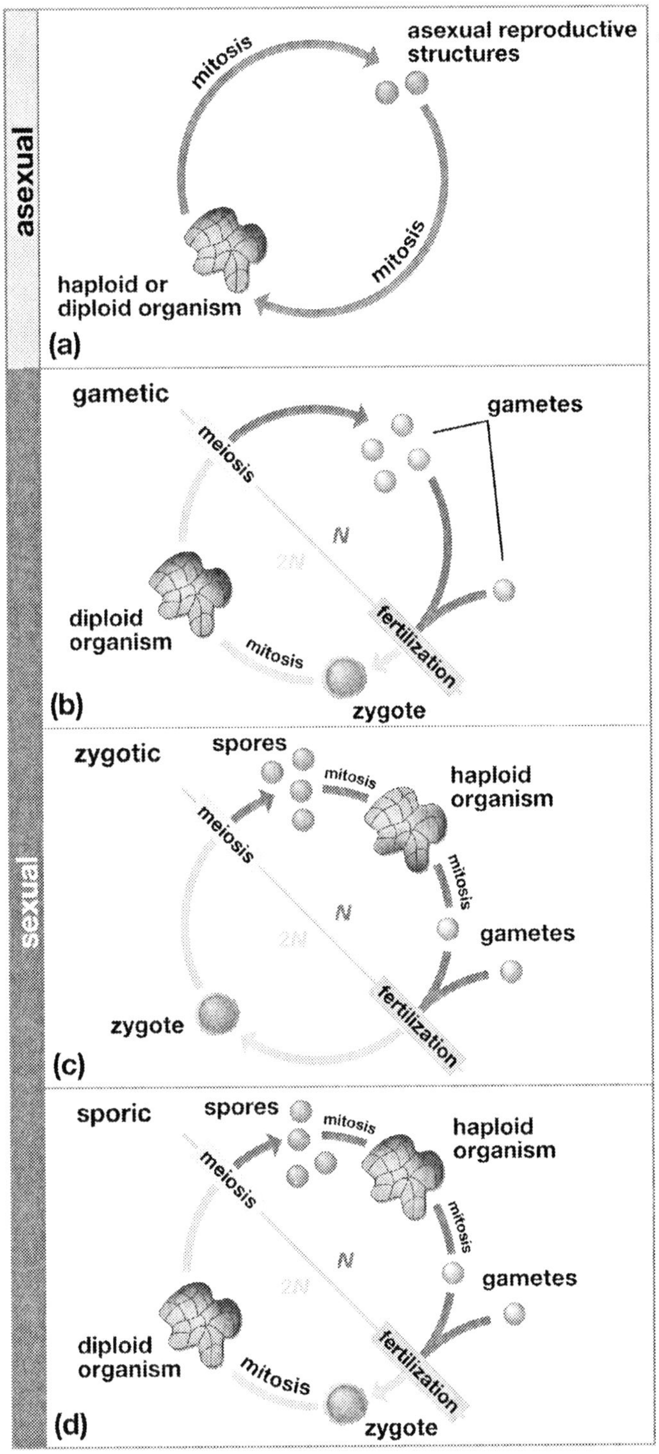

Figure 13.14 **A comparison of life cycles**

NOTES

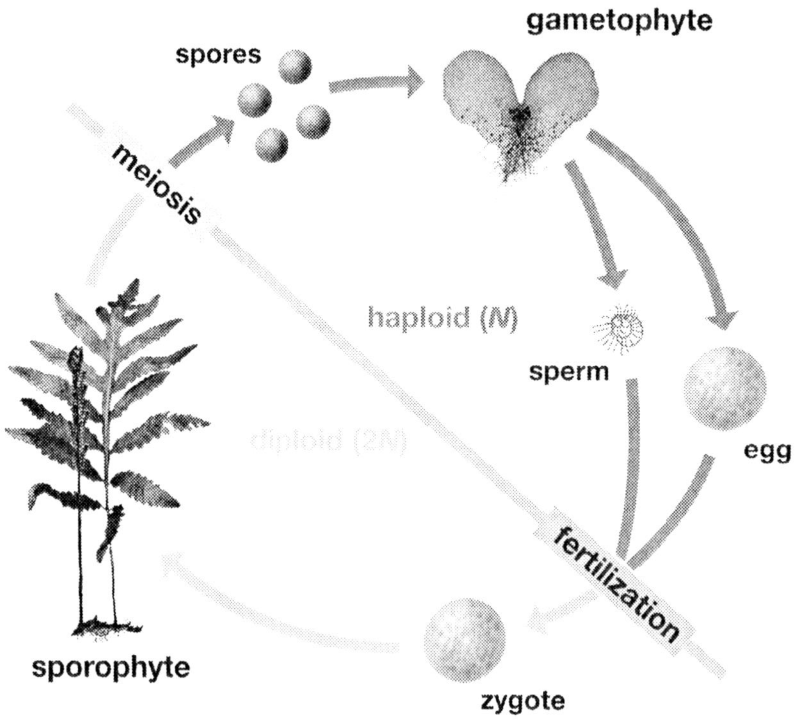

Figure 13.16 **Life cycle of a fern**

NOTES

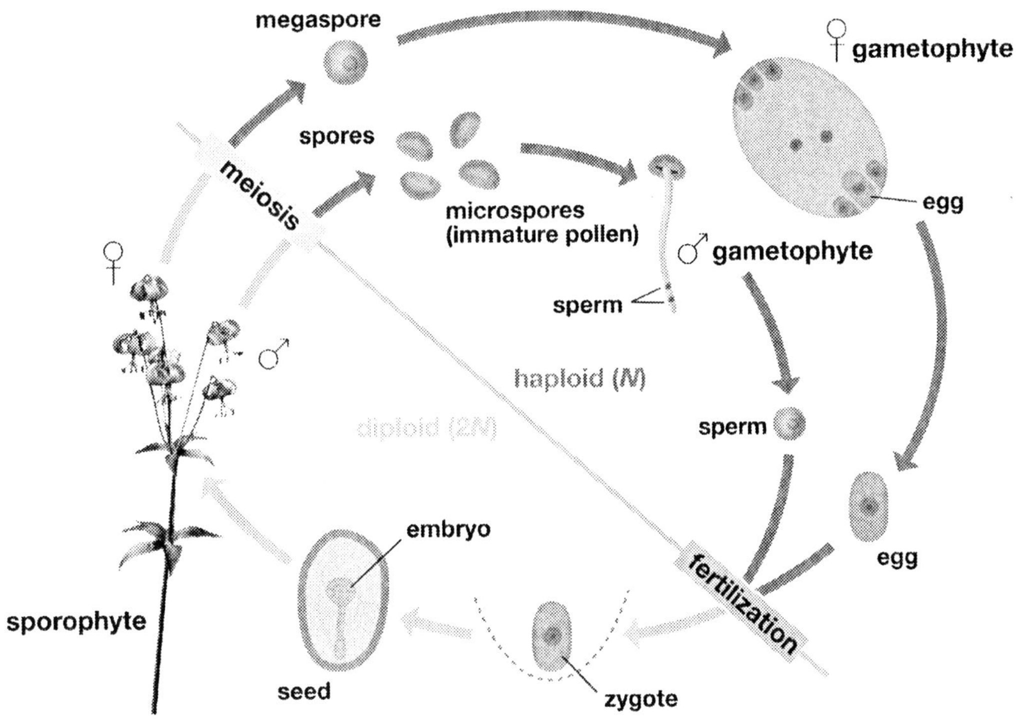

Figure 13.17 **Seed plant life cycle**

NOTES

NOTES

CHAPTER 14

Genetics and the Laws of Inheritance

Trait	Parental varieties		F_1 generation (Dominant trait)	F_2 generation		Ratio
				Dominant form	Recessive form	
Flower color	Purple	White	Purple	705 Purple	224 White	3.15:1
Seed color	Yellow	Green	Yellow	6022 Yellow	2001 Green	3.01:1
Seed shape	Round	Wrinkled	Round	5474 Round	1850 Wrinkled	2.96:1
Pod color	Green	Yellow	Green	428 Green	152 Yellow	2.82:1
Pod shape	Round	Constricted	Round	882 Round	299 Constricted	2.95:1
Plant height	Tall	Dwarf	Tall	787 Tall	277 Dwarf	2.84:1
Flower position	Axial	Top	Axial	651 Axial	207 Top	3.14:1

Figure 14.3 **Mendel's experiments with seven individual traits in garden peas**

NOTES

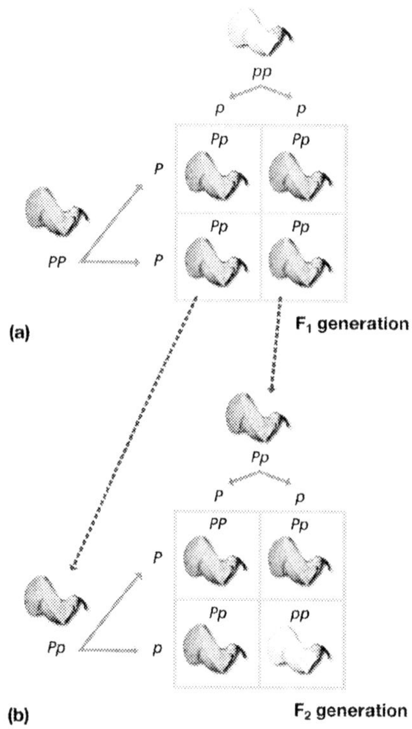

Figure 14.4 **Mendel's experiment with purple and white flowered peas**

NOTES

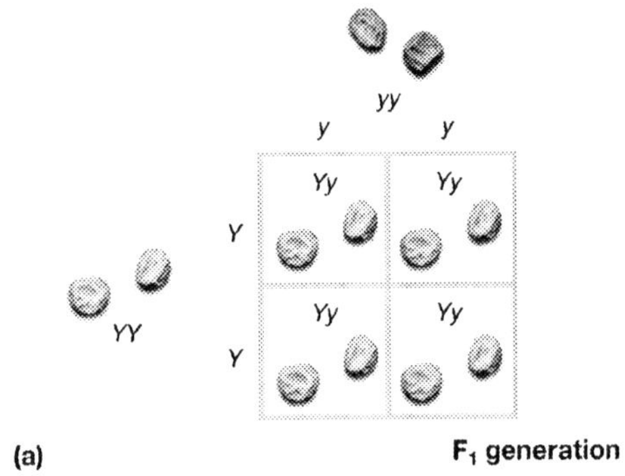

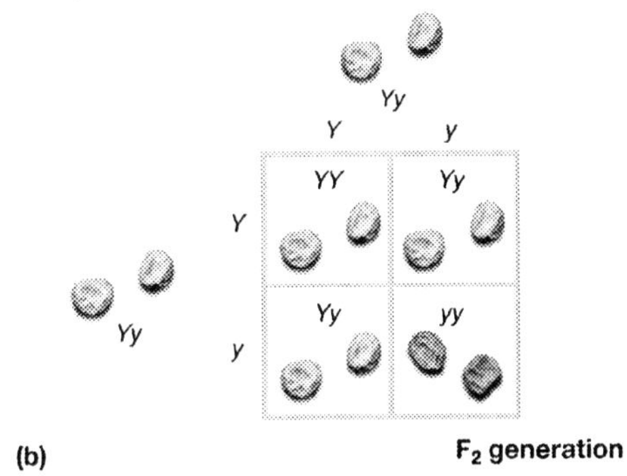

Figure 14.5 **Mendel's experiment with yellow and green seed color**

NOTES

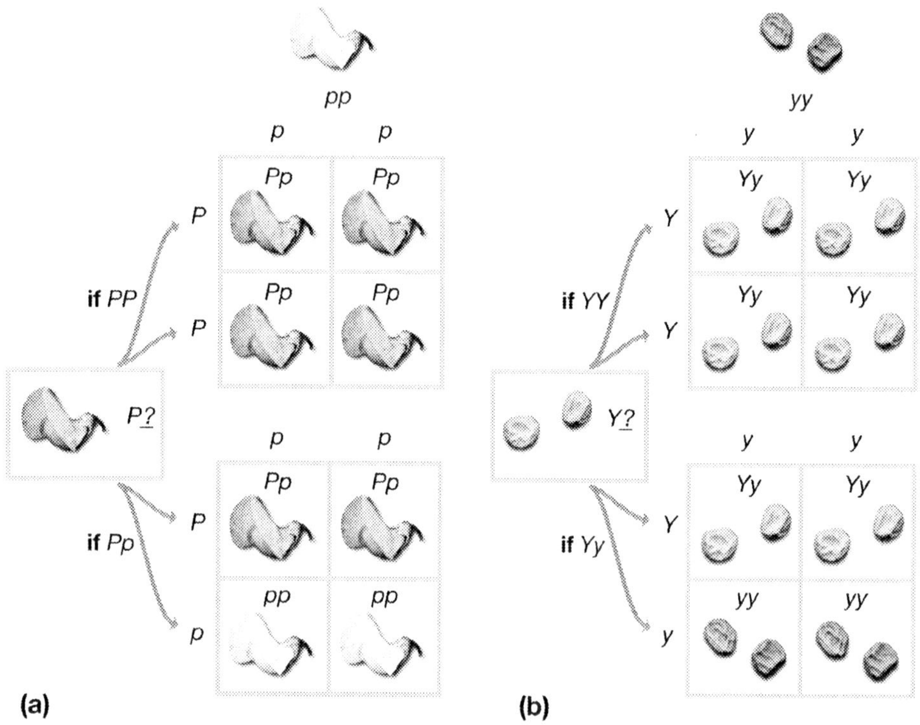

Figure 14.6 **Testcross for purple flower color and yellow seed color**

NOTES

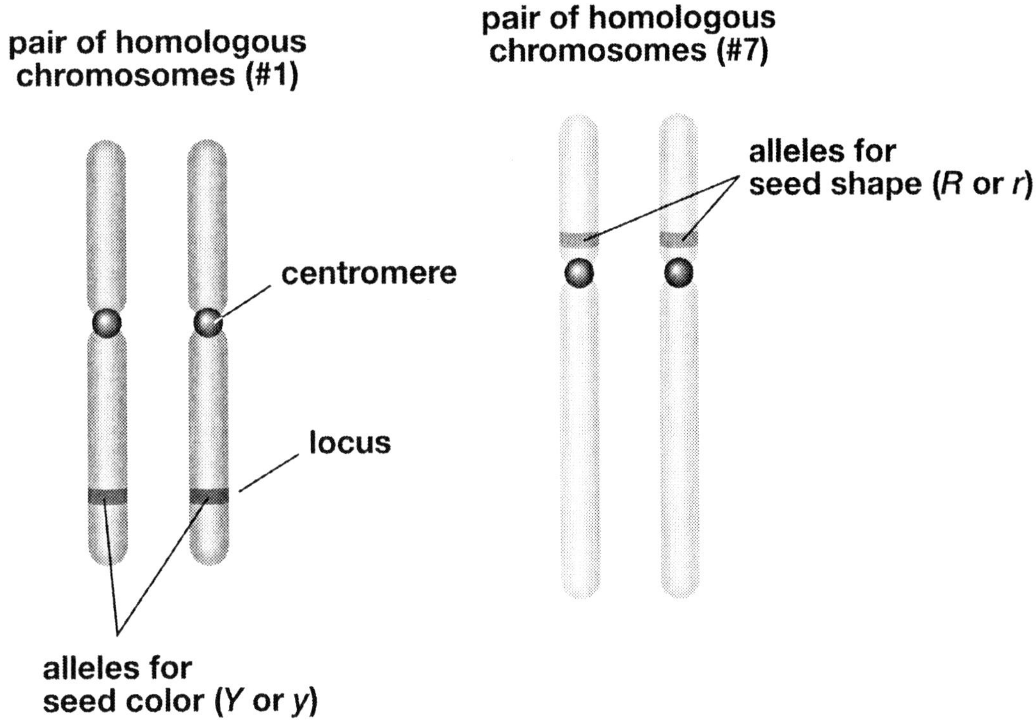

Figure 14.7 **Two homologous pairs of pea chromosomes**

NOTES

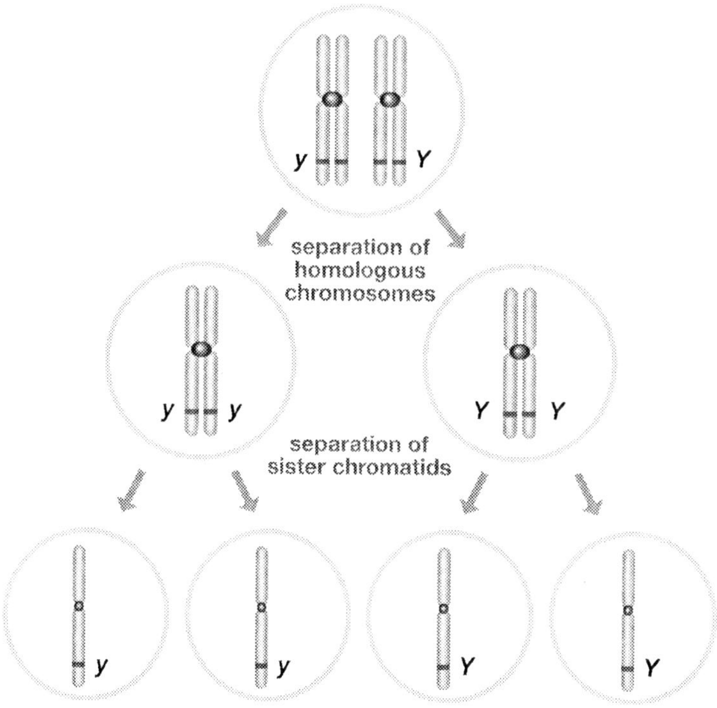

Figure 14.8 **Meiosis and the segregation of alleles**

NOTES

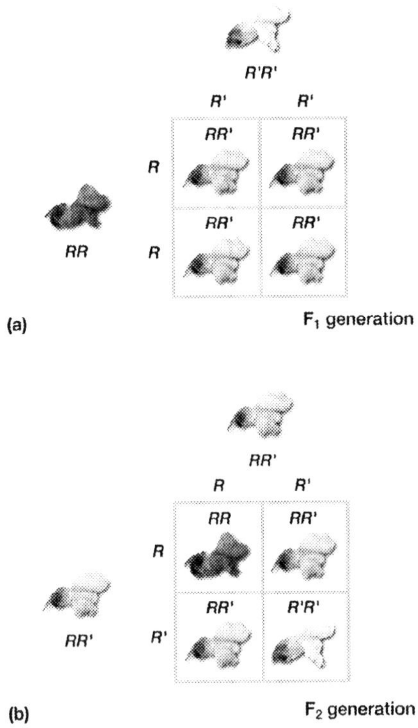

Figure 14.9 **Incomplete dominance in snapdragons**

NOTES

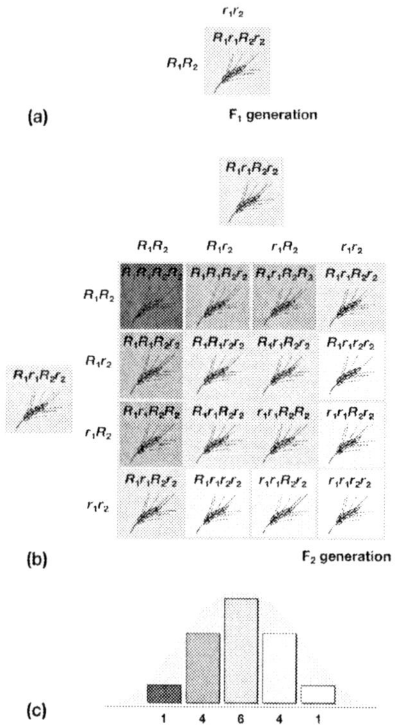

Figure 14.10 **Polygenic inheritance in wheat kernel color**

NOTES

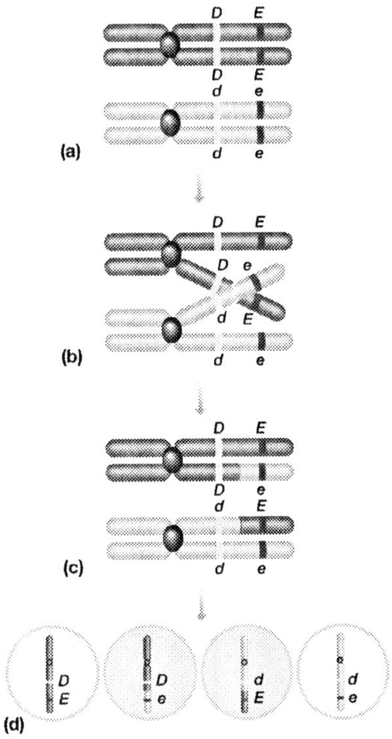

Figure 14.12 **Crossing over in the sweet pea, *Lathyrus odoratus***

NOTES

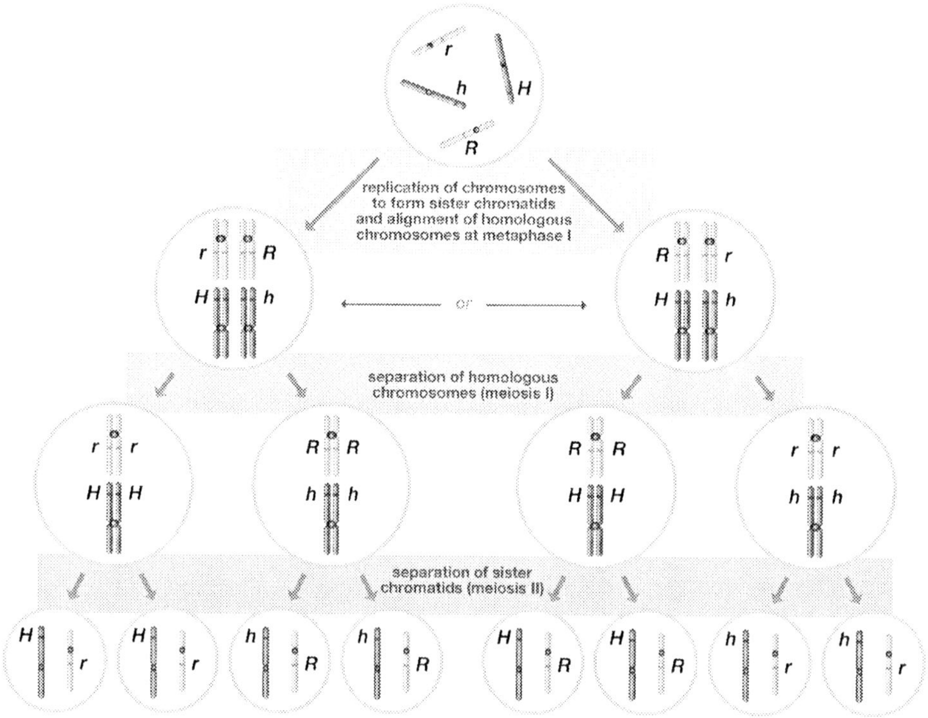

Figure 14.14 **Meiosis and the independent assortment of alleles**

NOTES

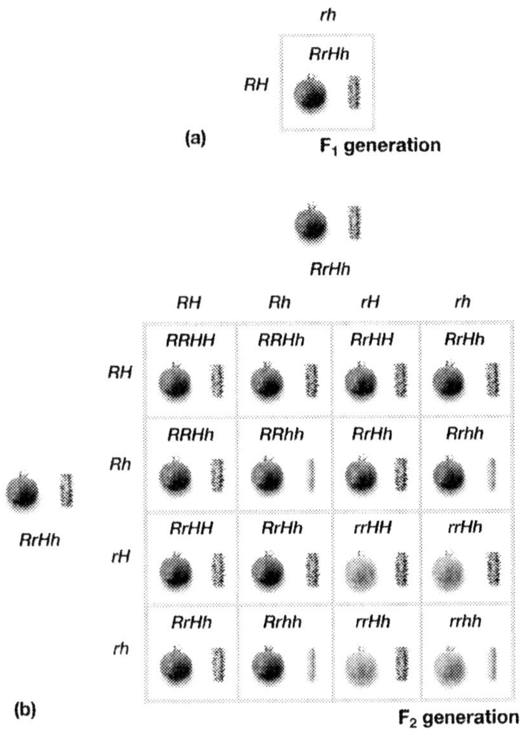

Figure 14.15 **Dihybrid cross in tomato**

NOTES

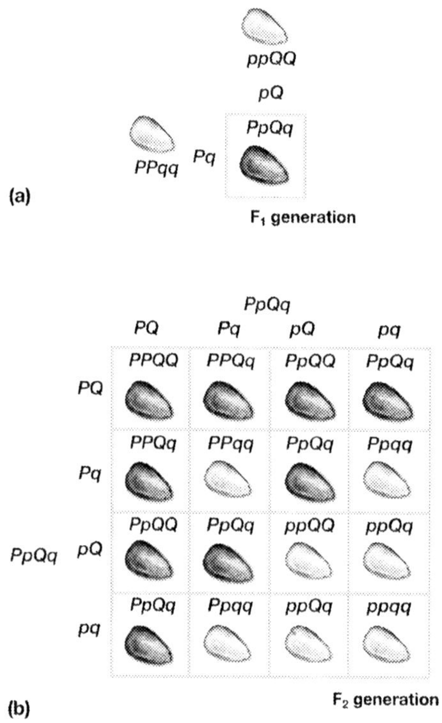

Figure 14.17 **Epistasis in corn for the genes for anthocyanin production**

NOTES

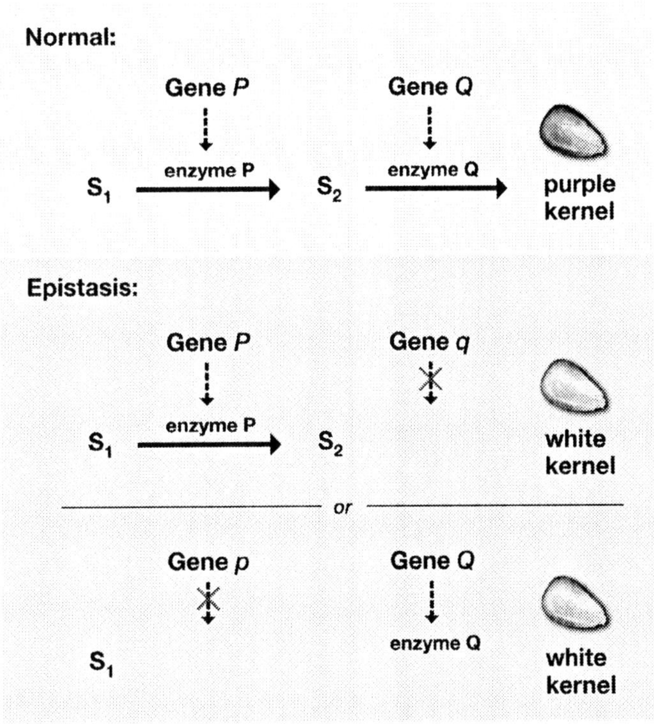

Figure 14.18 **Metabolic pathway for anthocyanin production**

NOTES

NOTES

CHAPTER 15

Genetic Engineering

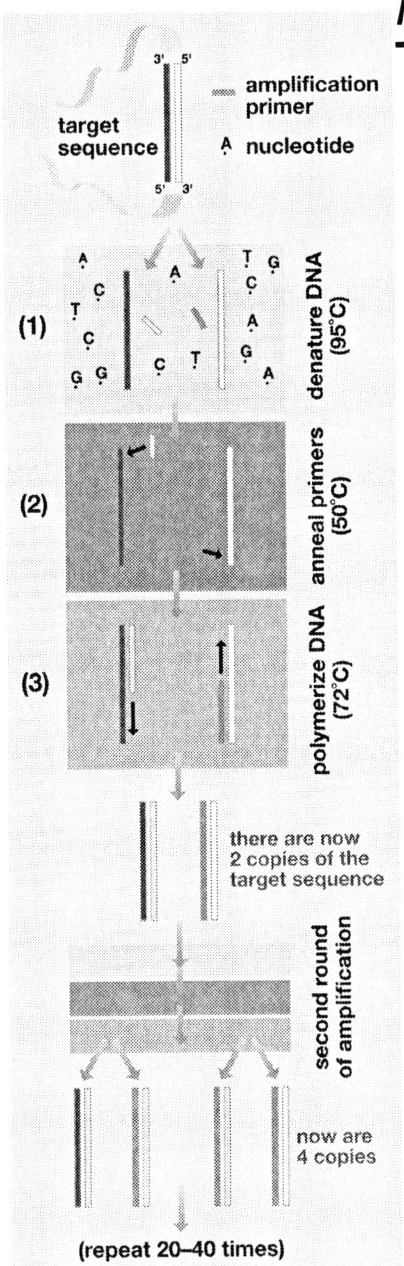

E15.1 **Polymerase chain reaction**

NOTES

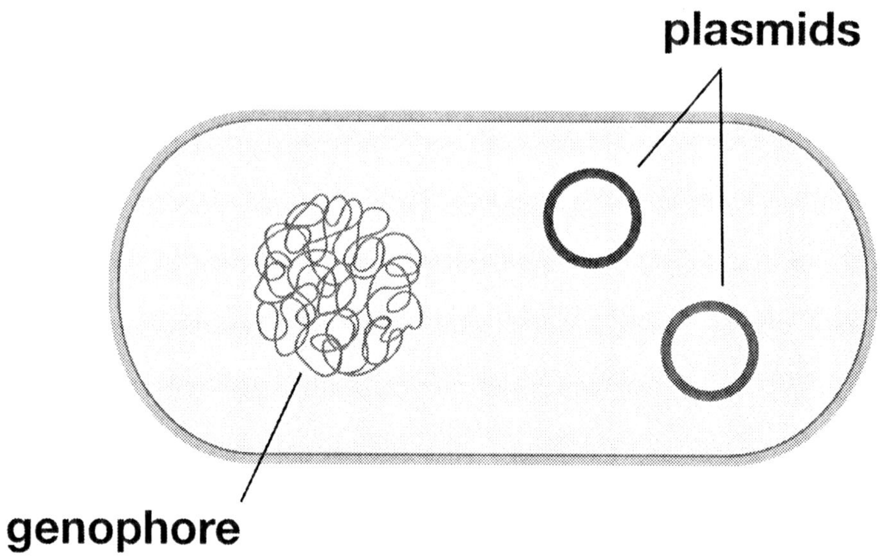

Figure 15.6 **Bacterial cell, its genophore, and plasmids**

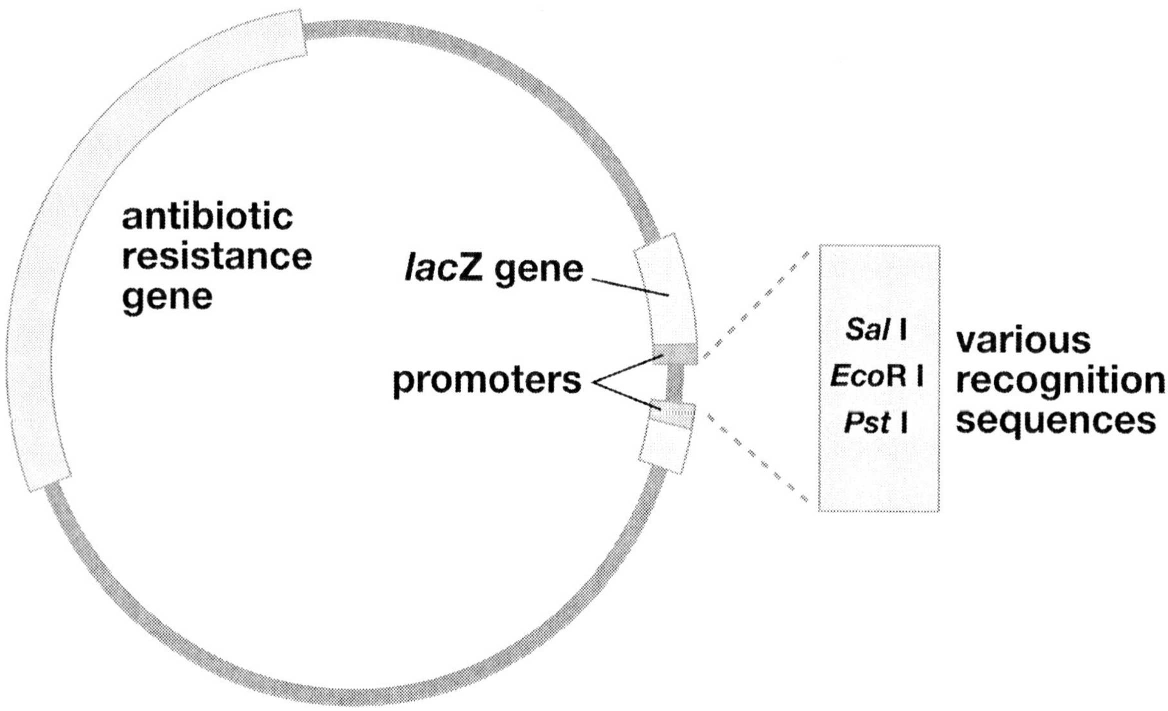

Figure 15.7 **Simplified diagram of plasmid with antibiotic resistance gene marker**

NOTES

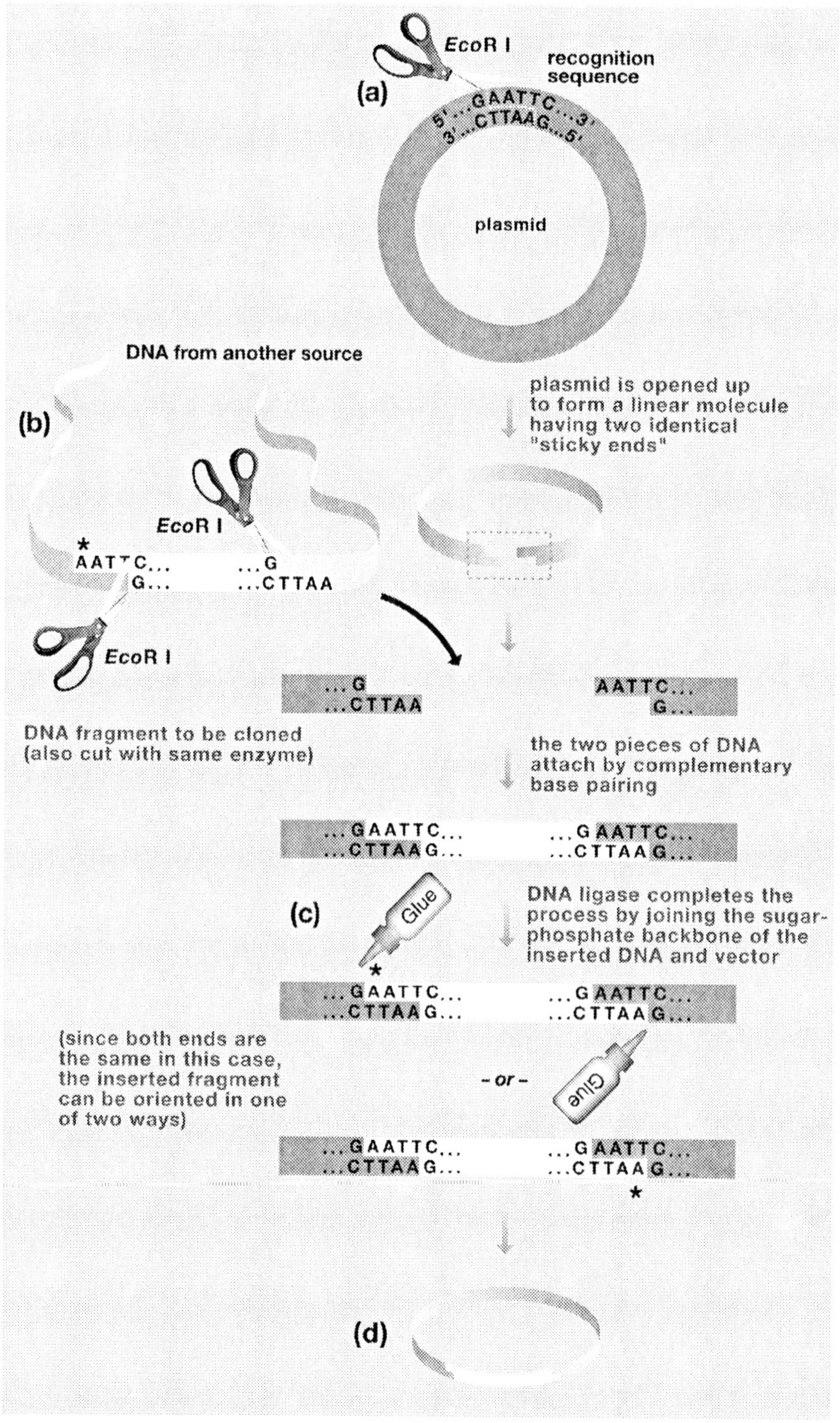

Figure 15.8 **Restricting the plasmid and DNA to be cloned and joining the two together**

NOTES

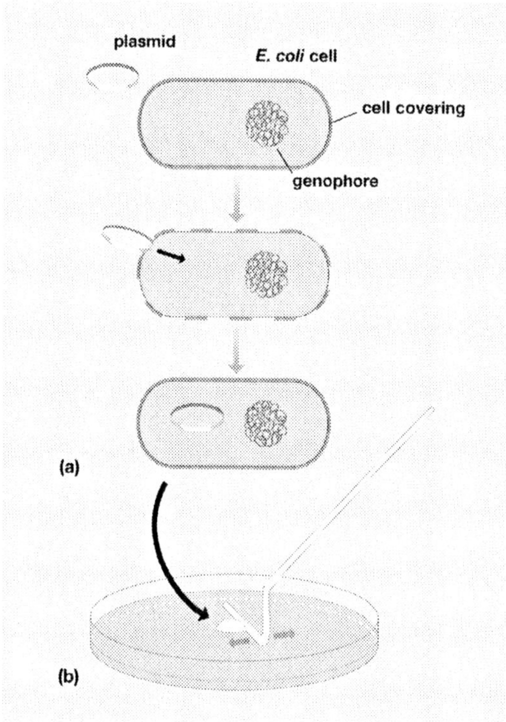

Figure 15.9 **Insertion of plasmids into bacterial cells—transformation**

NOTES

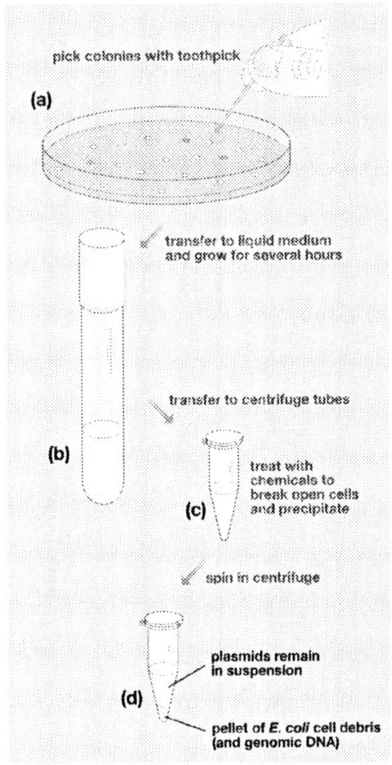

Figure 15.10 **Harvesting plasmids from transformed cells for further manipulation**

NOTES

NOTES

CHAPTER 16

Biological Evolution

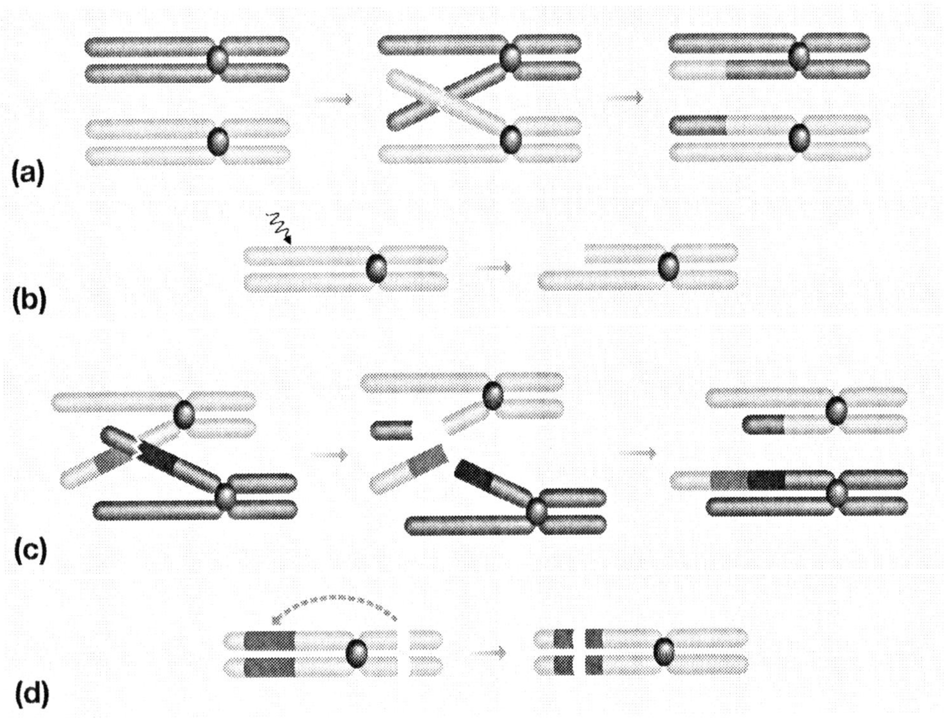

Figure 16.13 **Chromosome mutations**

NOTES

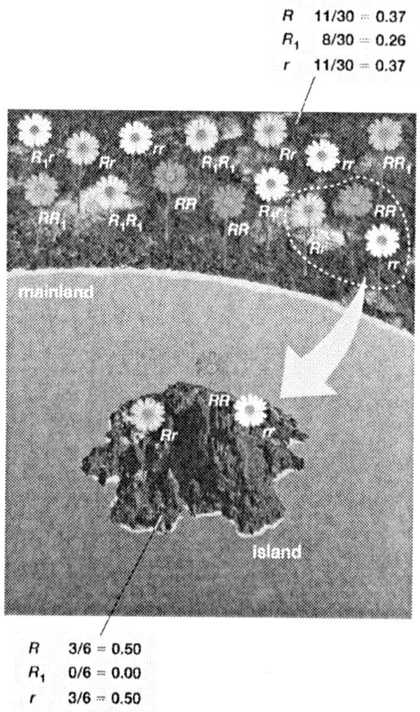

Figure 16.14 **The founder effect**

NOTES

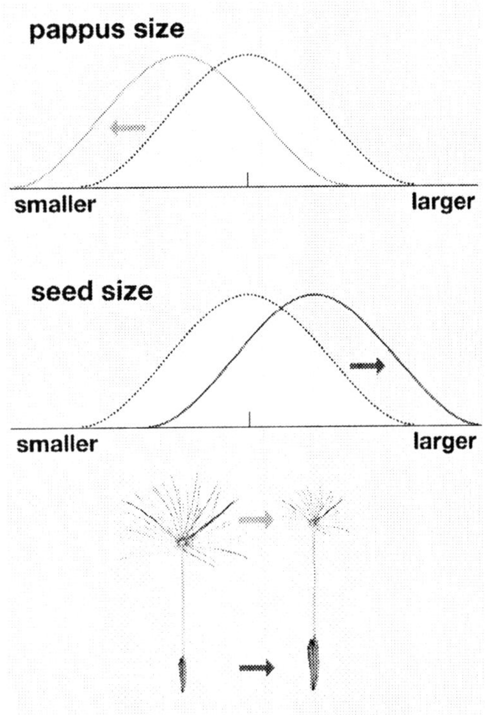

Figure 16.15 **Directional selection**

NOTES

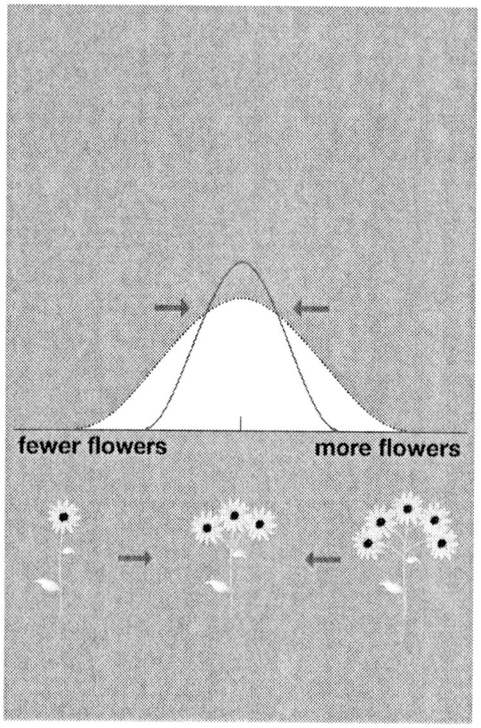

Figure 16.16 **Stabilizing selection**

NOTES

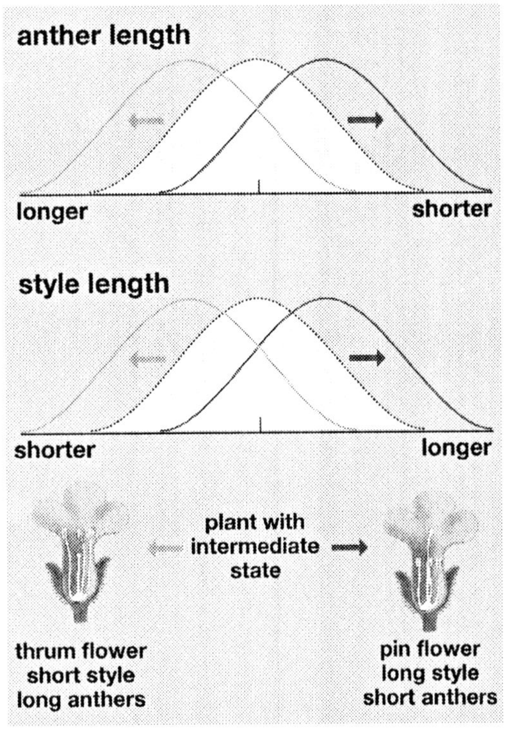

Figure 16.17 **Disruptive selection**

NOTES

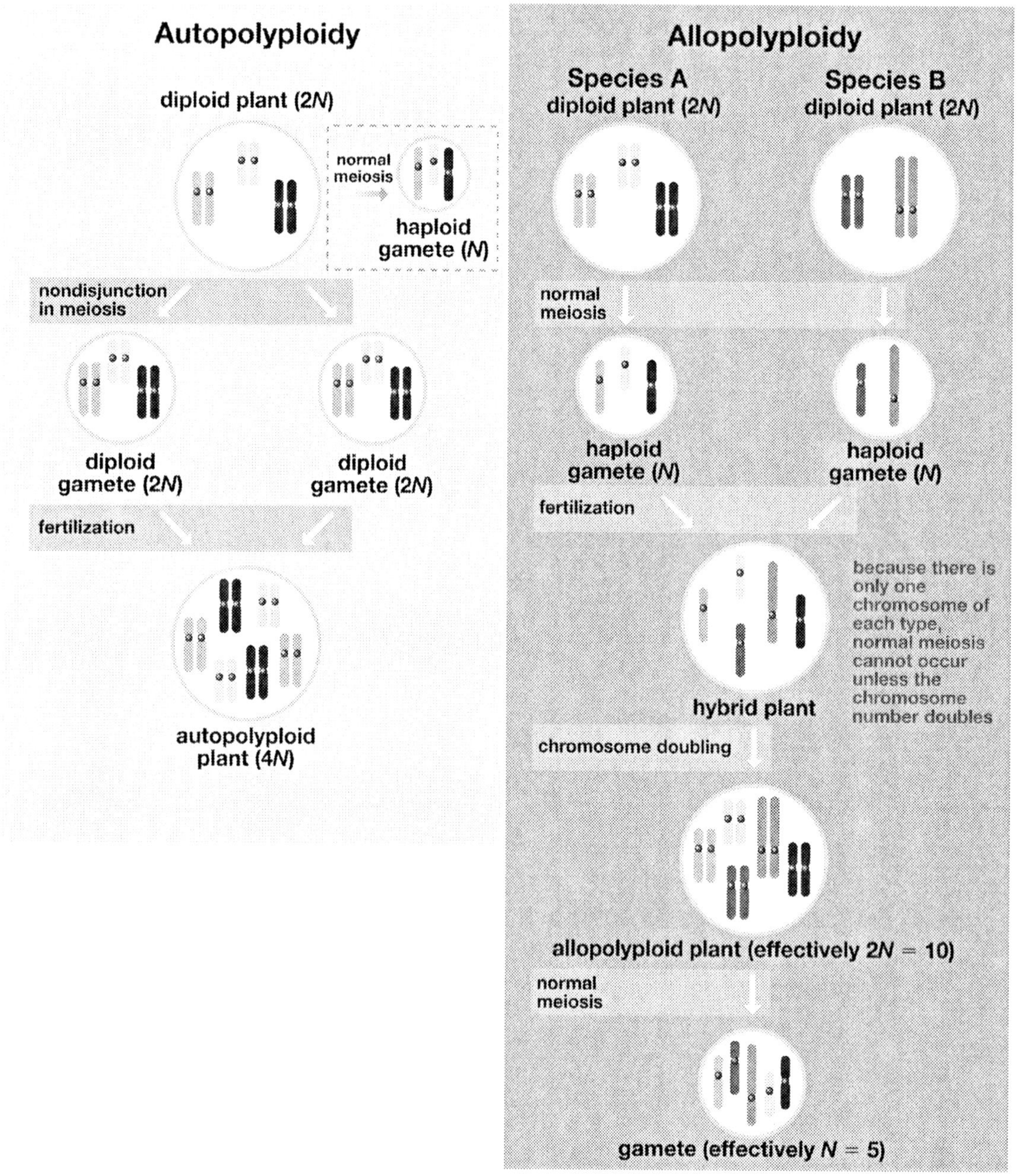

Figure 16.20 **Autopolyploidy and allopolyploidy**

NOTES

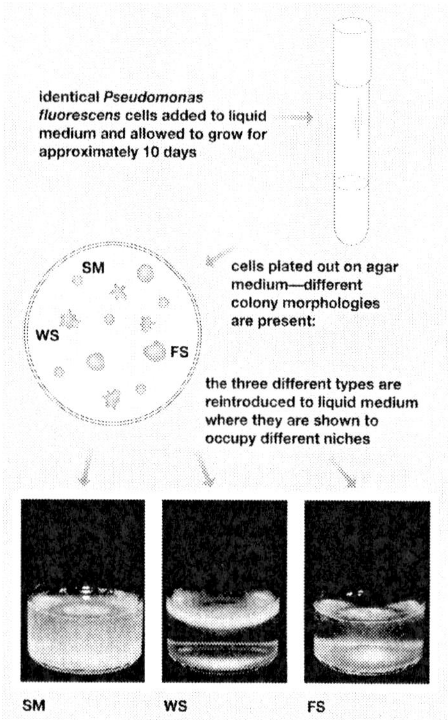

Figure 16.22 **Rapid development of microbial diversity in vials of sugar water**

NOTES

CHAPTER 17

Naming and Organizing Plants and Microbes

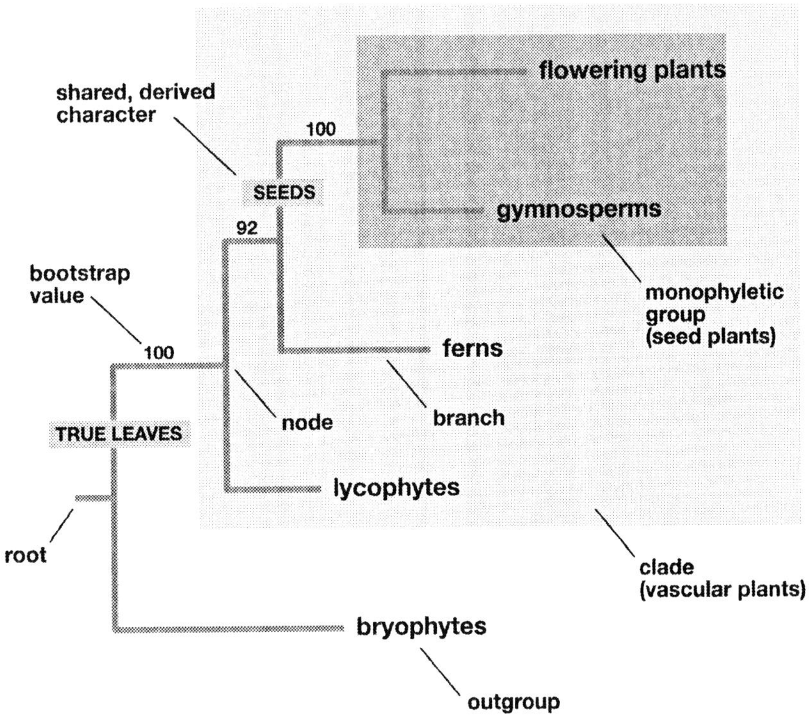

Figure 17.15 **An example of a phylogenetic tree illustrating the relationships of land plant groups**

NOTES

NOTES

CHAPTER 18

Prokaryotes and the Origin of Life on Earth

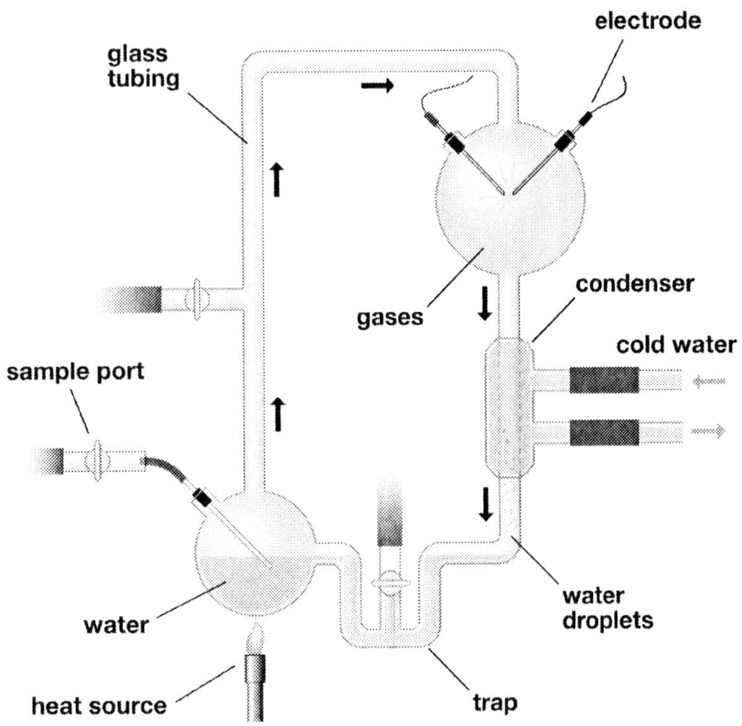

Figure 18.6 **Urey-Miller experimental apparatus**

NOTES

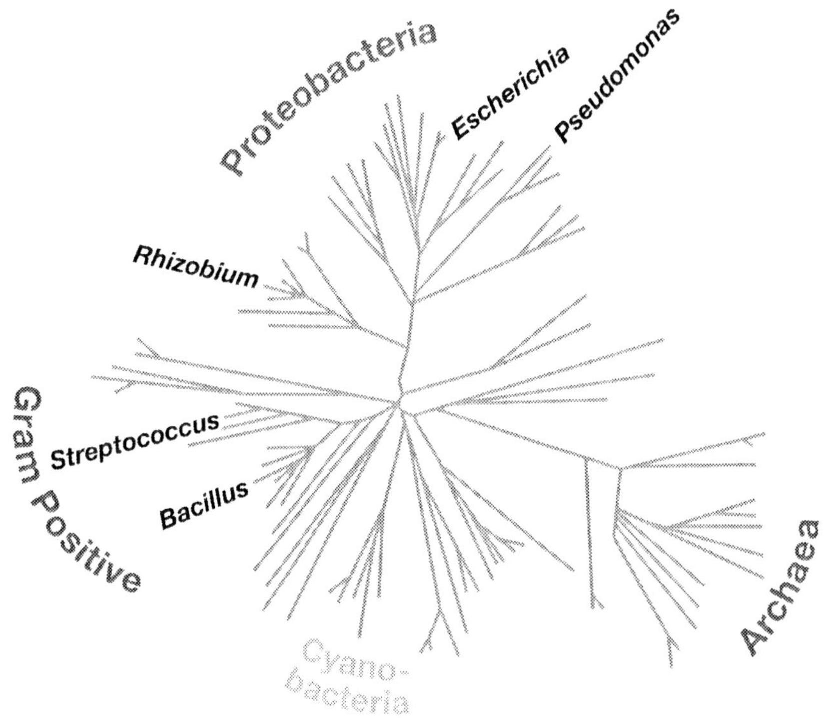

Figure 18.19 **Phylogenetic tree showing major groups of prokaryotes**

NOTES

CHAPTER 19

Protists and the Origin of Eukaryotic Cells

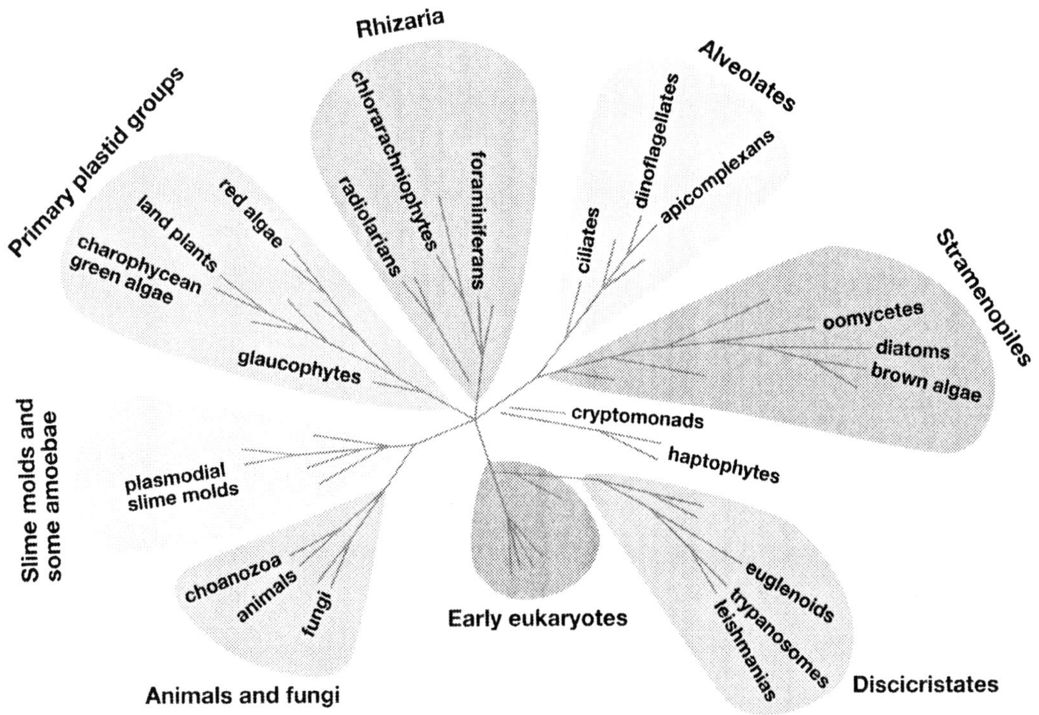

Figure 19.8 **Evolutionary relationships of algae and other protists**

NOTES

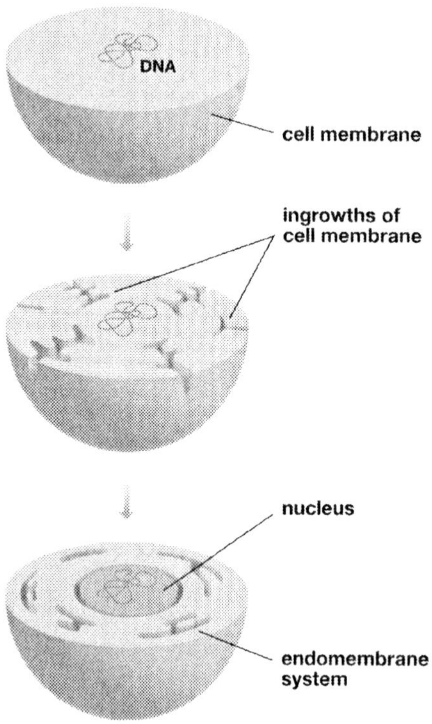

Figure 19.13 **The eukaryotic nucleus and endomembrane system are thought to have originated by infolding of the cell membrane in an ancestor of modern eukaryotic cells**

NOTES

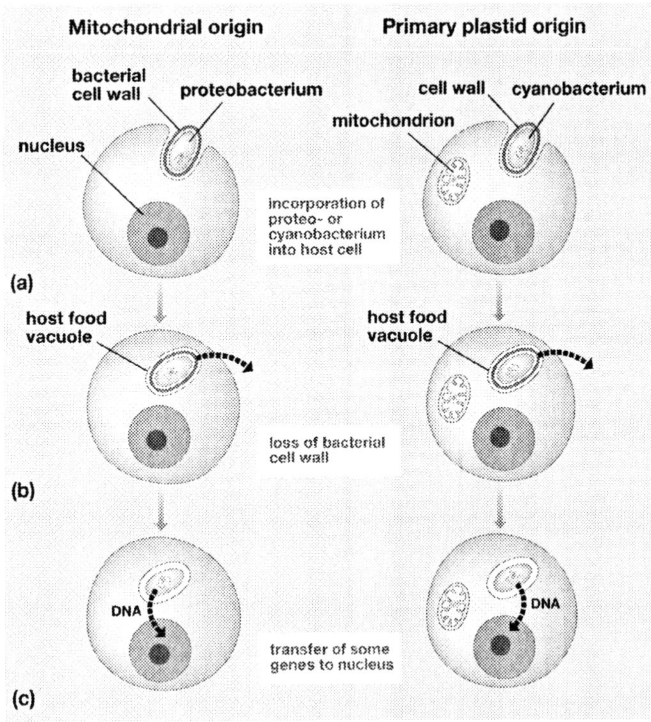

Figure 19.15 **Origin of mitochondria and primary plastids by endosymbiosis**

NOTES

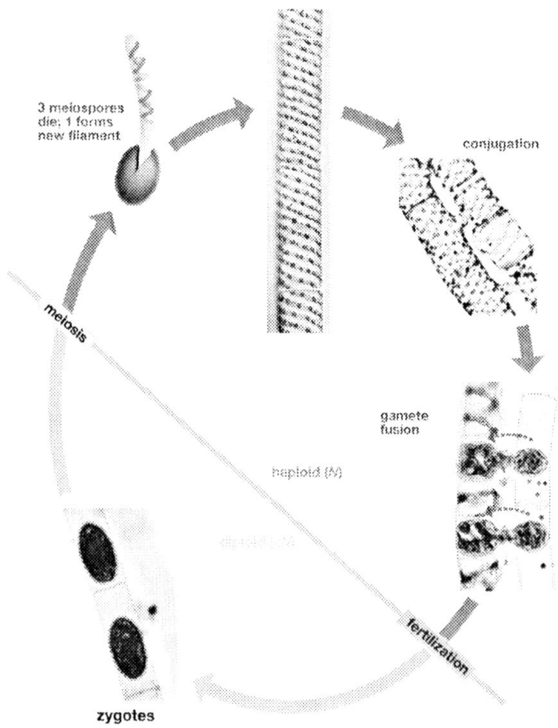

Figure 19.20 **The green algae *Spirogyra* has a zygotic life cycle**

NOTES

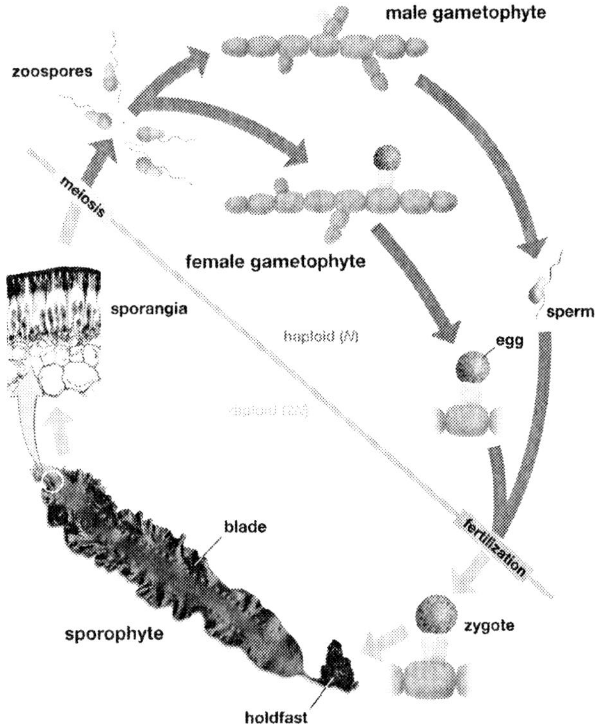

Figure 19.21 **The brown (stramenopile) alga *Laminaria* has a sporic life cycle involving alternation of distinct multicellular sporophyte and gametophyte generations**

NOTES

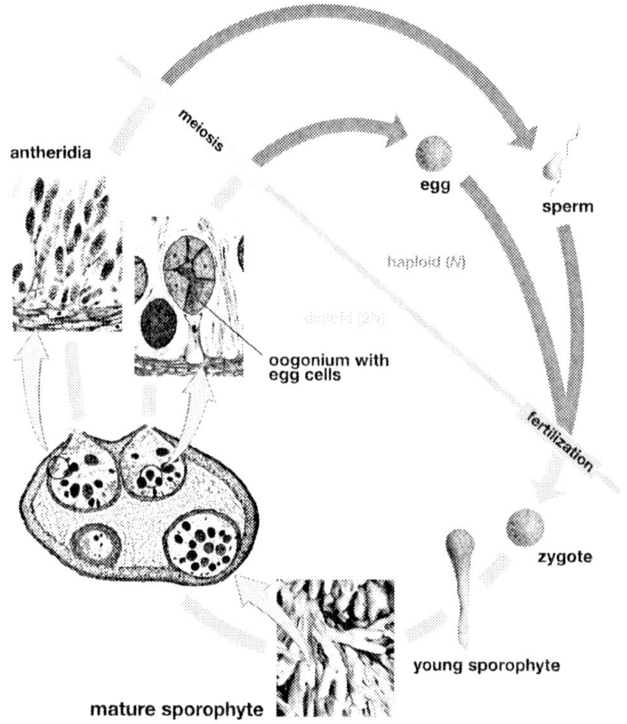

Figure 19.22 **The rockweed *Fucus* has a gametic life cycle similar to that of animals**

NOTES

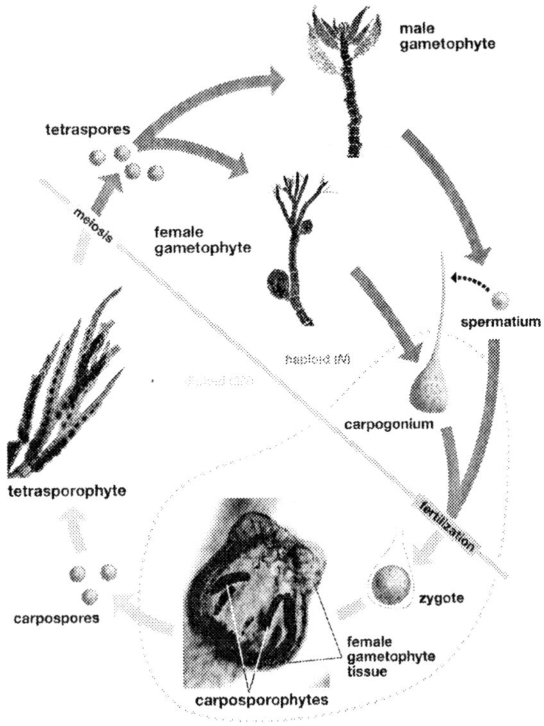

Figure 19.23 **Most red algae have a life cycle that includes alternation of three distinct multicellular generations**

NOTES

NOTES

CHAPTER 20

Fungi and Lichens

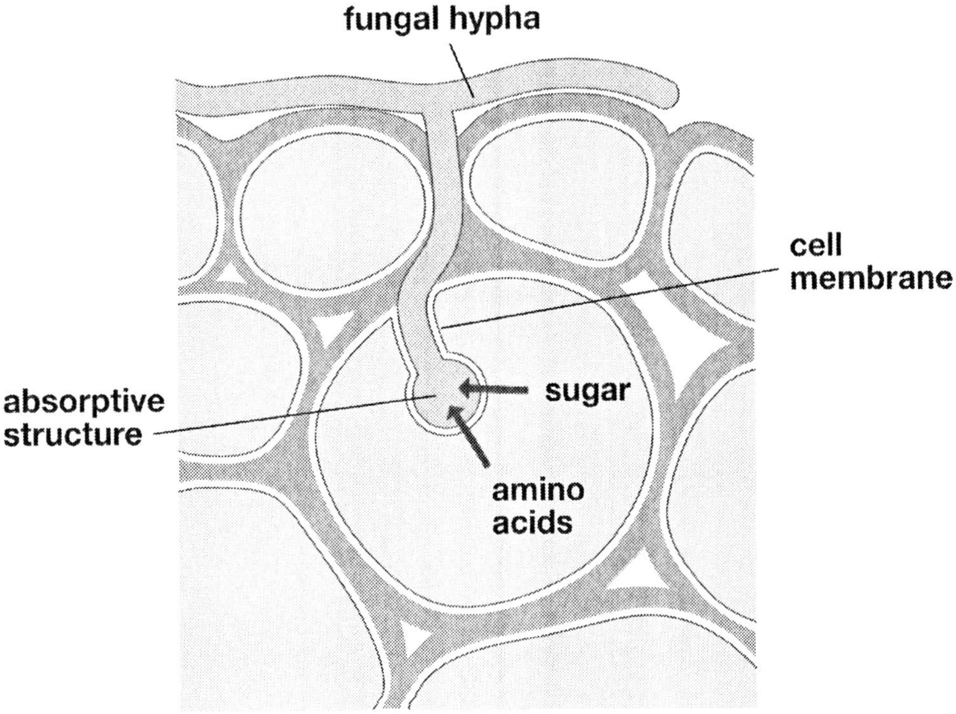

Figure 20.4 **Fungal pathogens of plants**

NOTES

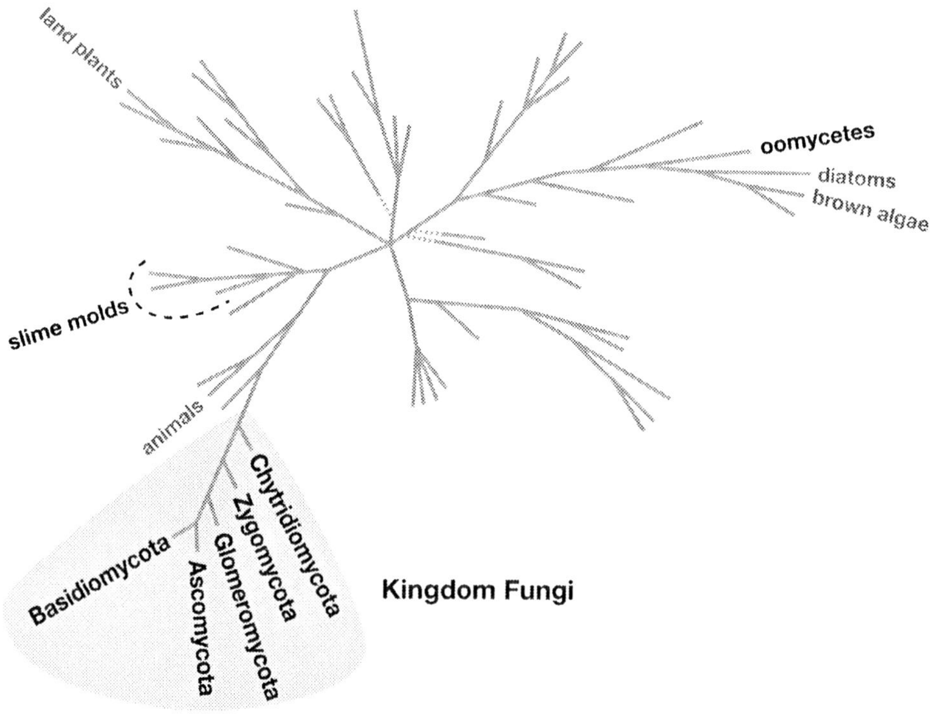

Figure 20.6 **Relationships of the sexually reproducing groups of fungi**

NOTES

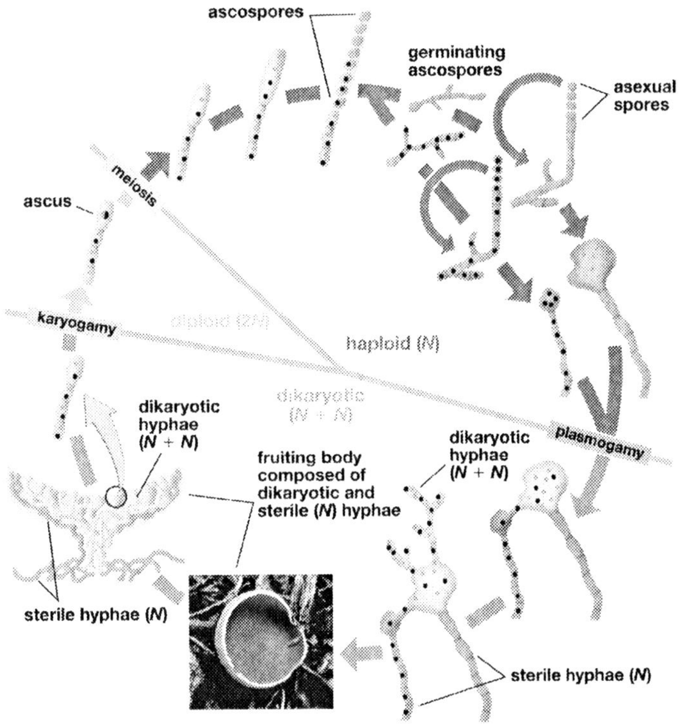

Figure 20.10 **Ascomycete life cycle**

NOTES

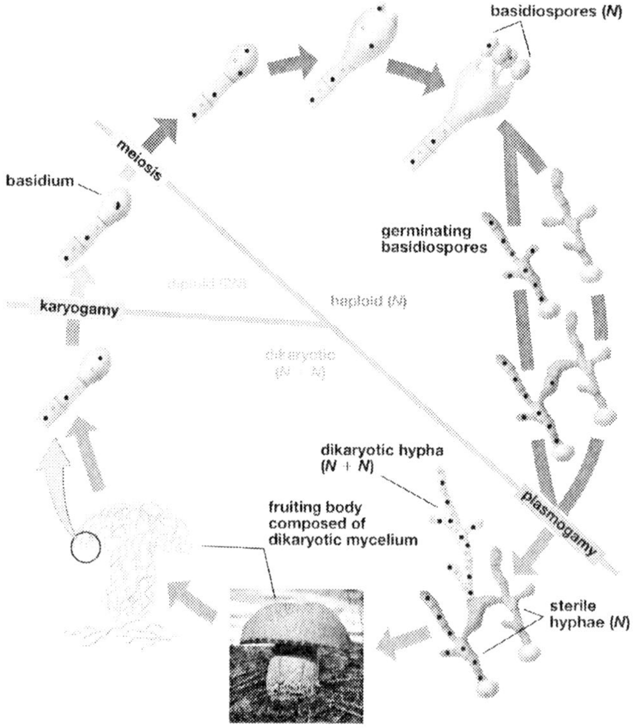

Figure 20.12 **Basidiomycete life cycle**

NOTES

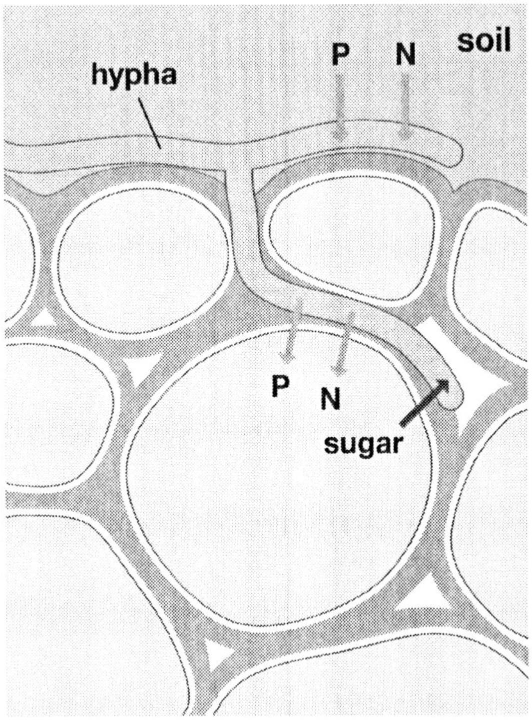

Figure 20.17 **Echotomycorrhizae**

NOTES

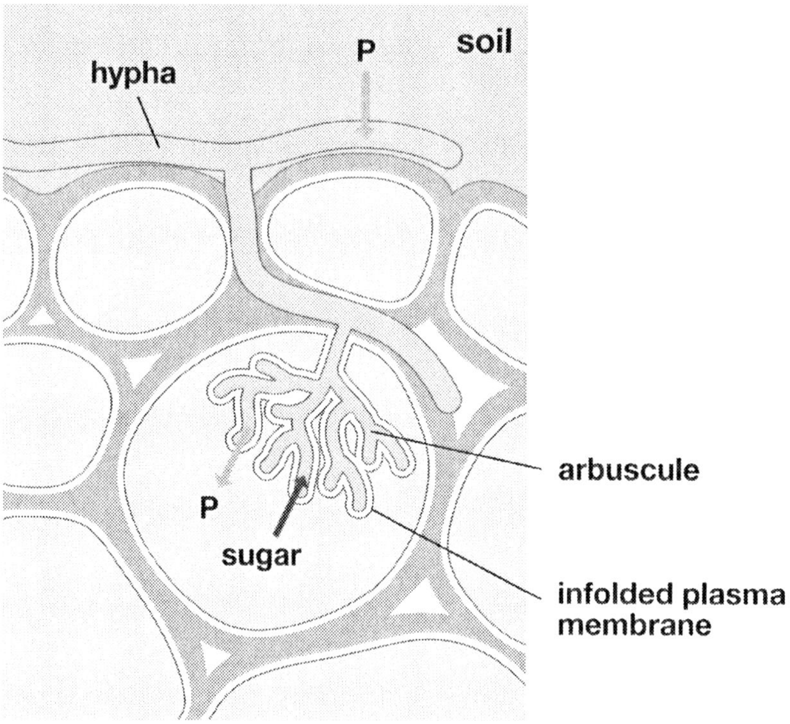

Figure 20.18 **Endomycorrhizae**

NOTES

CHAPTER 21 Seedless Plants

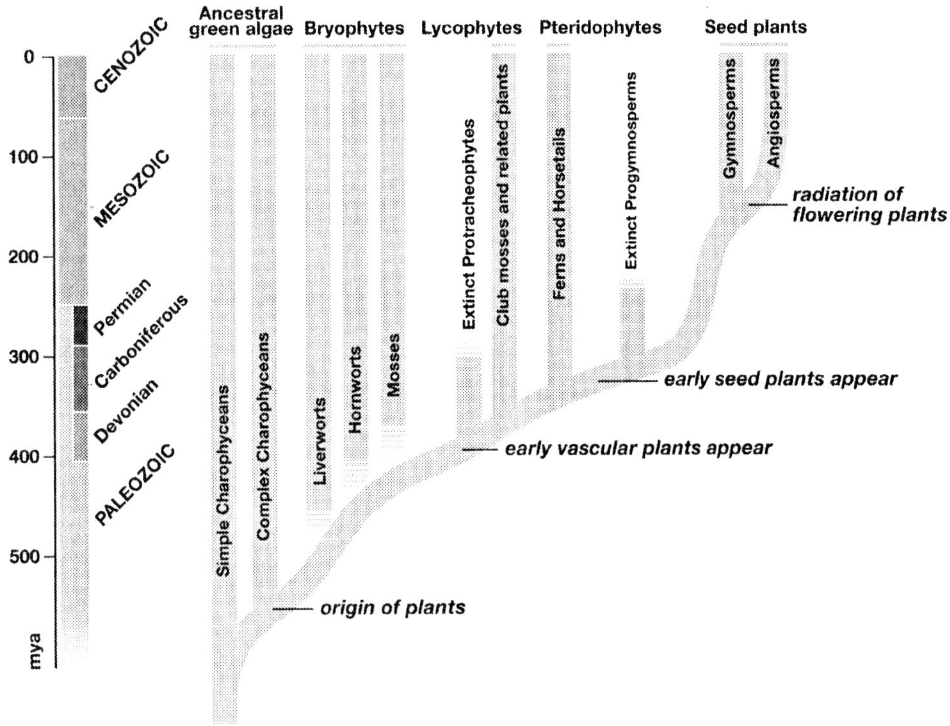

Figure 21.3 **Diagram of plant phylogeny on a geological timescale**

NOTES

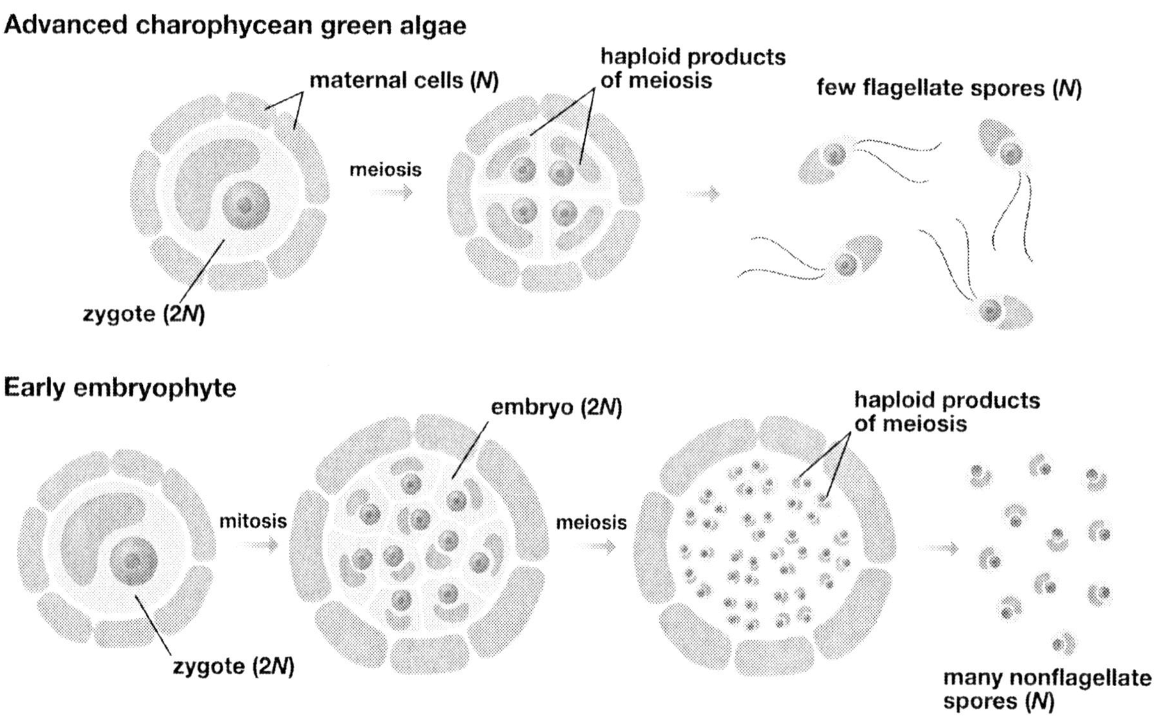

Figure 21.6 **Origin of the plant embryo and sporophyte**

NOTES

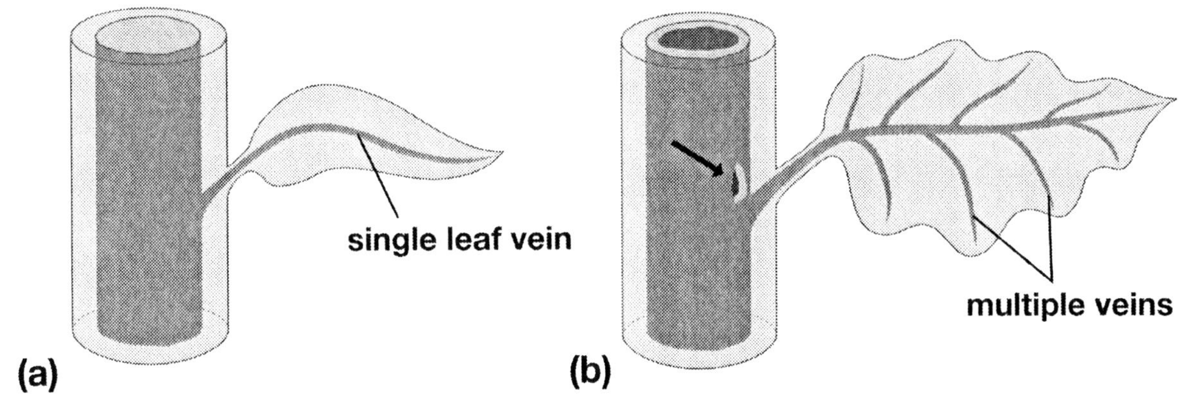

Figure 21.7 **Lycophylls versus euphylls**

NOTES

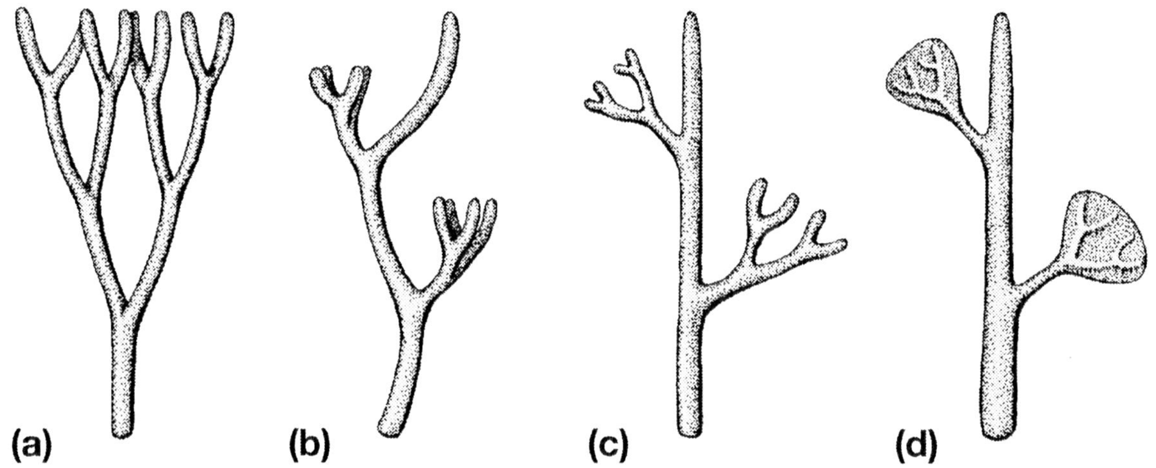

Figure 21.8 **Leaf evolution**

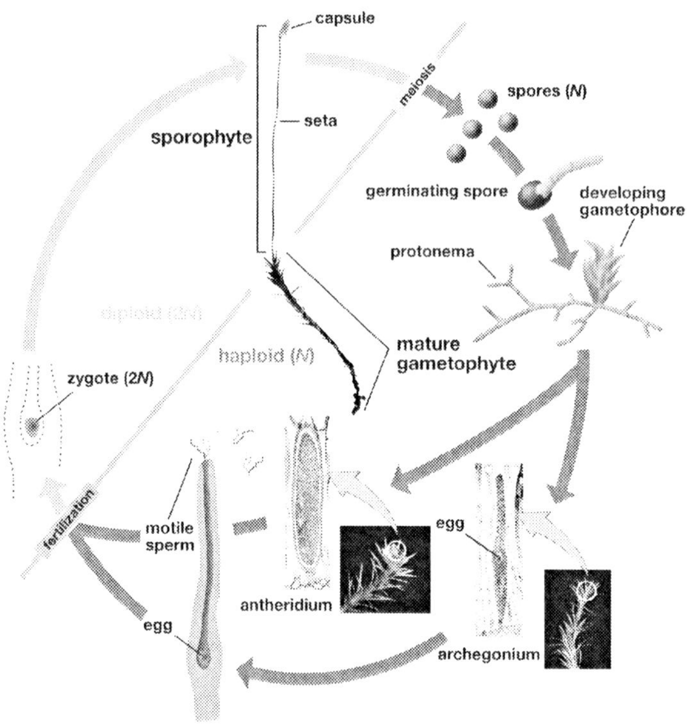

Figure 21.12 **Bryophyte life cycle**

NOTES

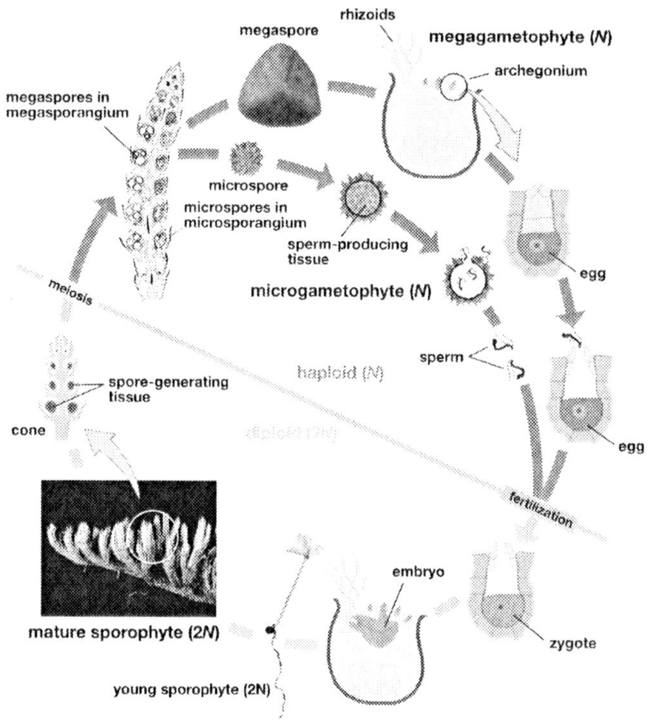

Figure 21.20 **Life history of a lycophyte**

NOTES

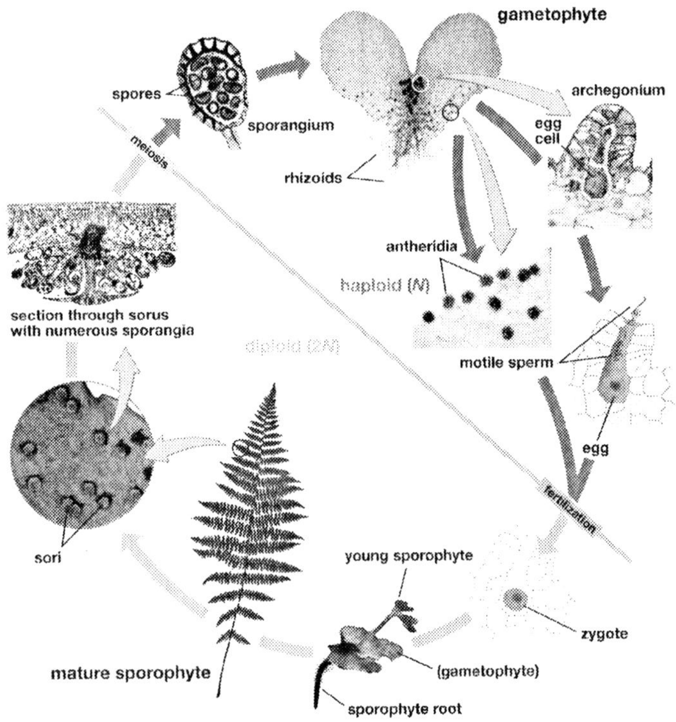

Figure 21.25 **Life cycle of a homosporous fern**

NOTES

NOTES

CHAPTER 22 Gymnosperms, the First Seed Plants

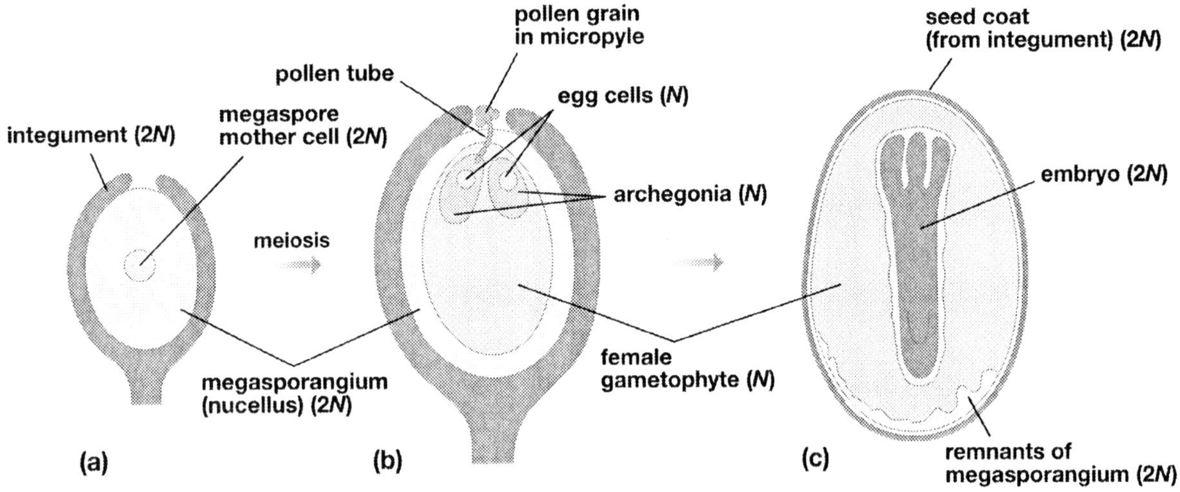

Figure 22.5 **Comparison of ovule and seed structure**

NOTES

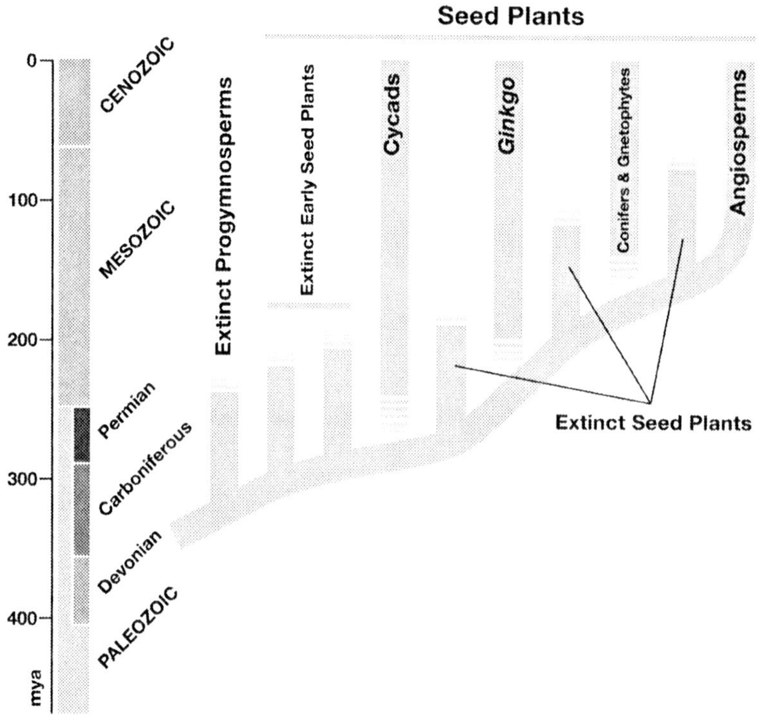

Figure 22.6 **Phylogeny of seed plants on a geological timescale**

NOTES

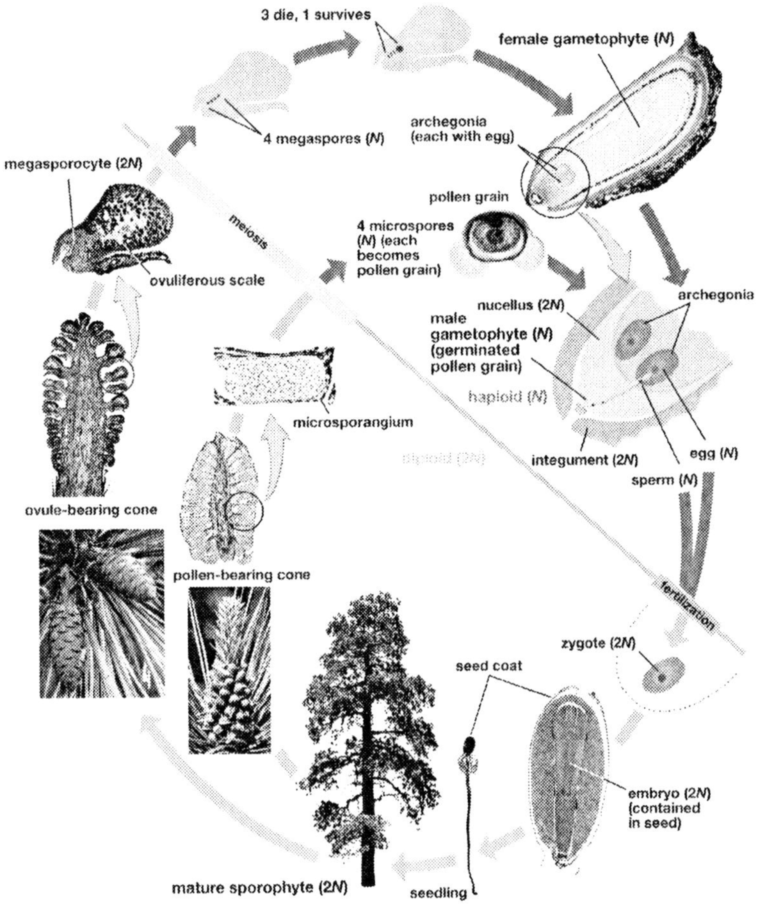

Figure 22.14 **Pine life cycle**

NOTES

NOTES

CHAPTER 23

Angiosperm Diversity and Reproduction

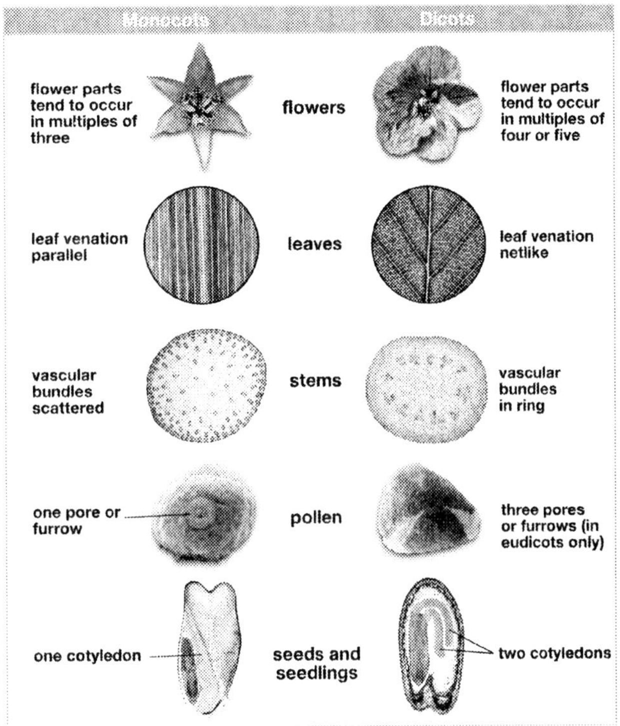

Figure 23..3 **Characteristics of monocots and dicots**

NOTES

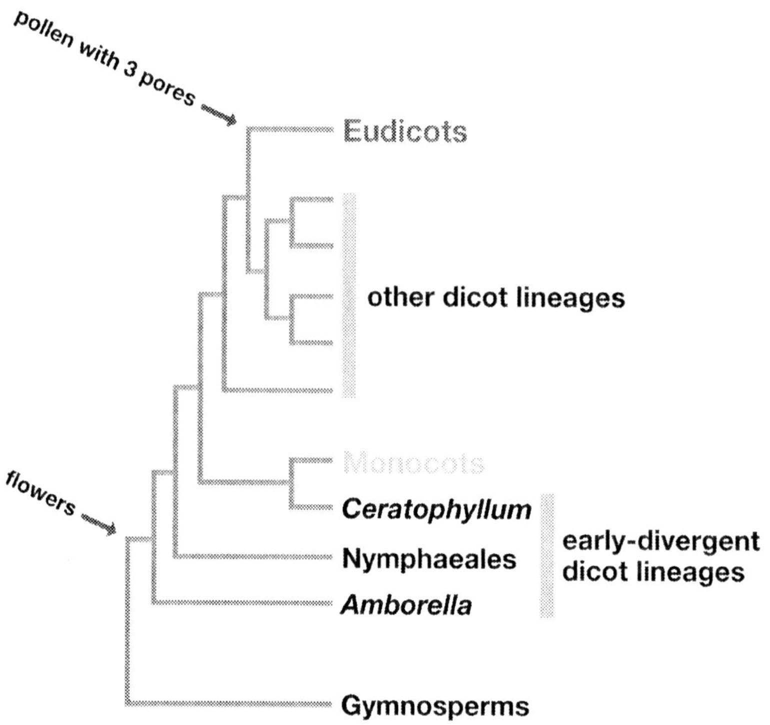

Figure 23.4 **Phylogenetic relationships among the major groups of flowering plants**

NOTES

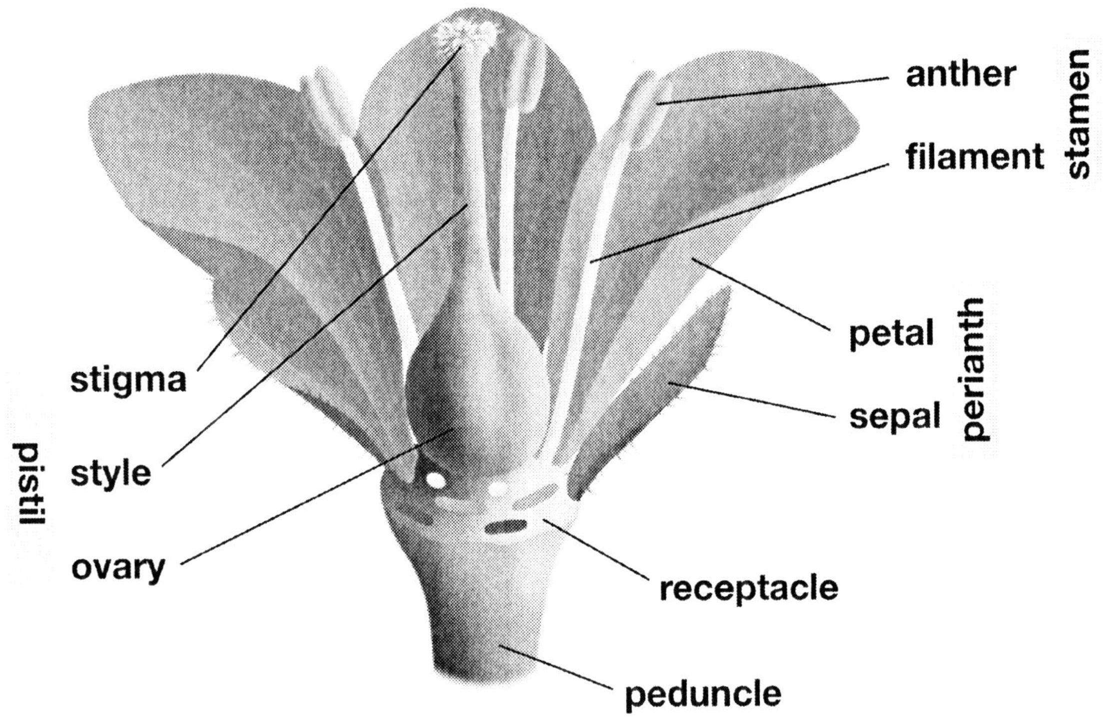

Figure 23.6 **Parts of a typical flower**

NOTES

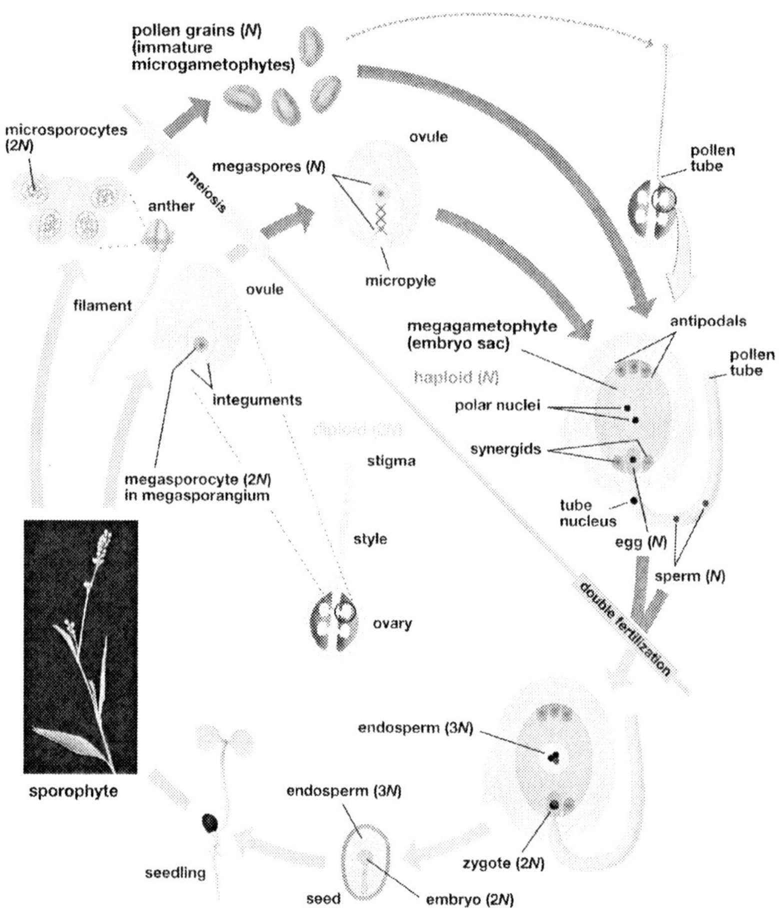

Figure 23.14 **Life cycle of a flowering plant**

NOTES

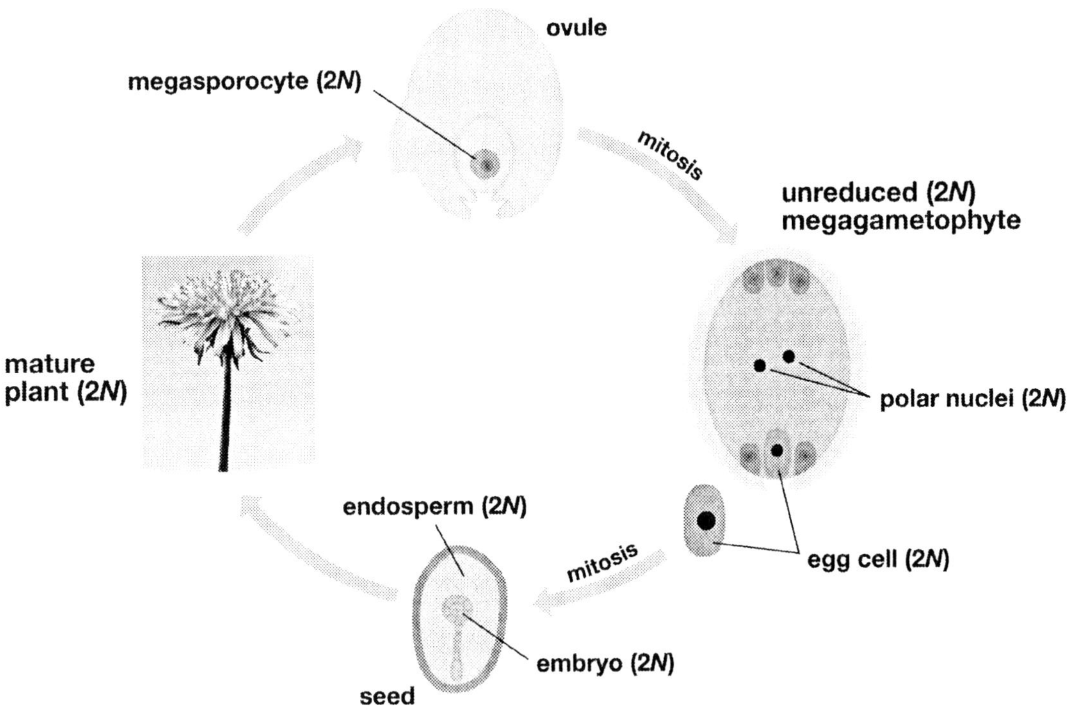

Figure 23.16 **Apomixis by diplospory**

NOTES

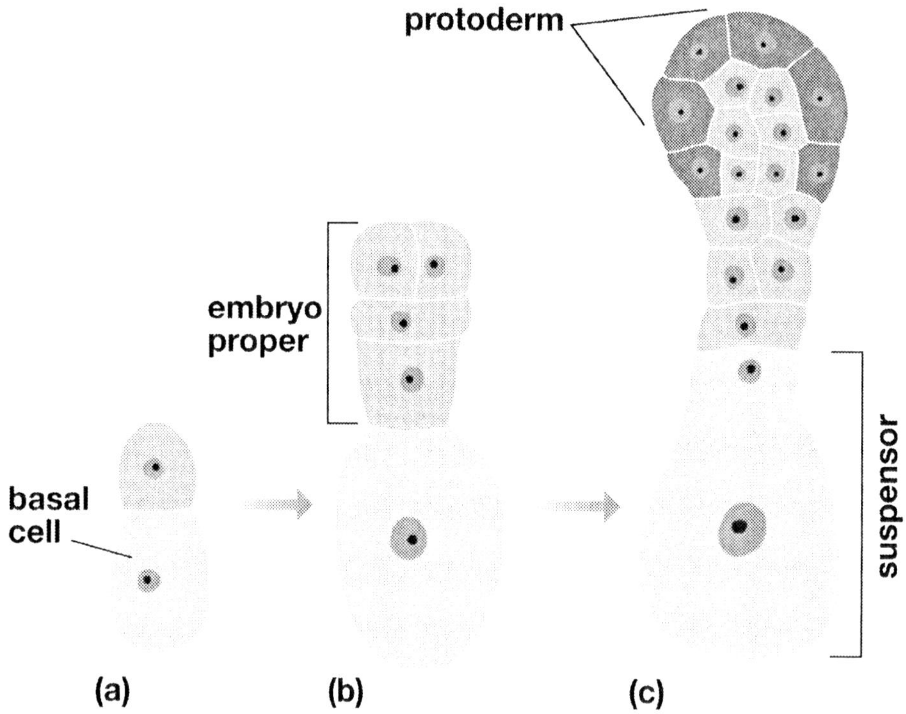

Figure 23.17 **Early stages in the development of the angiosperm embryo**

NOTES

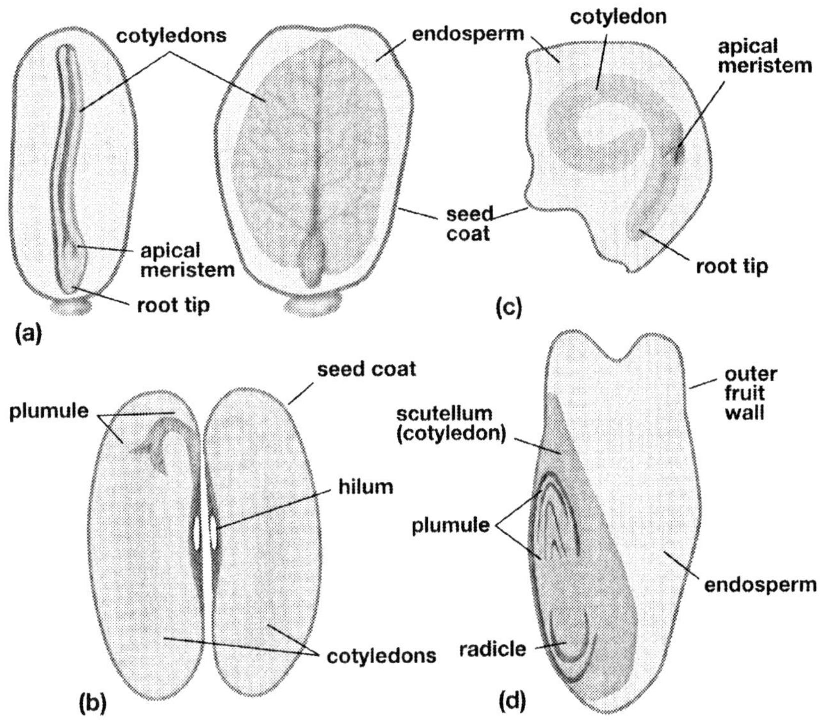

Figure 23.19 **Seeds of common eudicots and monocots**

NOTES

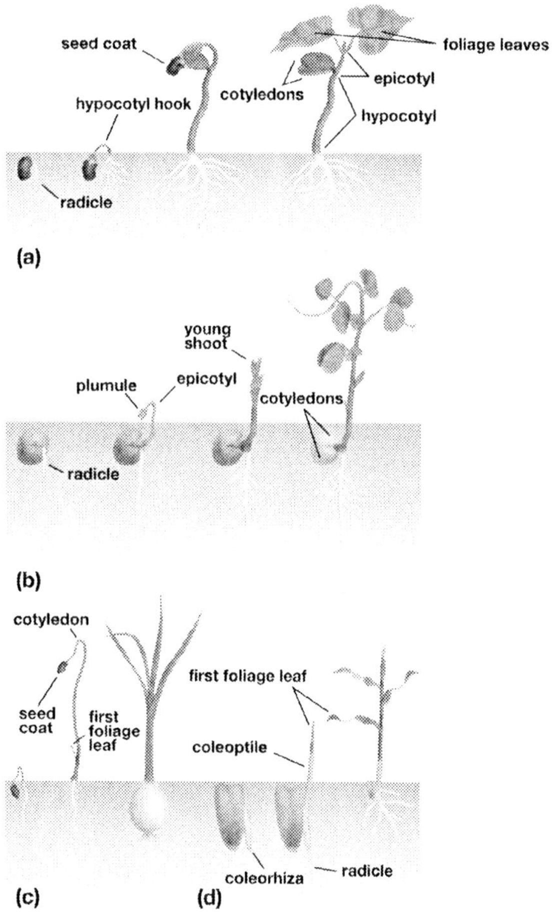

Figure 23.24 **Epigeous and hypogeous seed germination**

NOTES

CHAPTER 24 Flowering Plant and Animal Coevolution

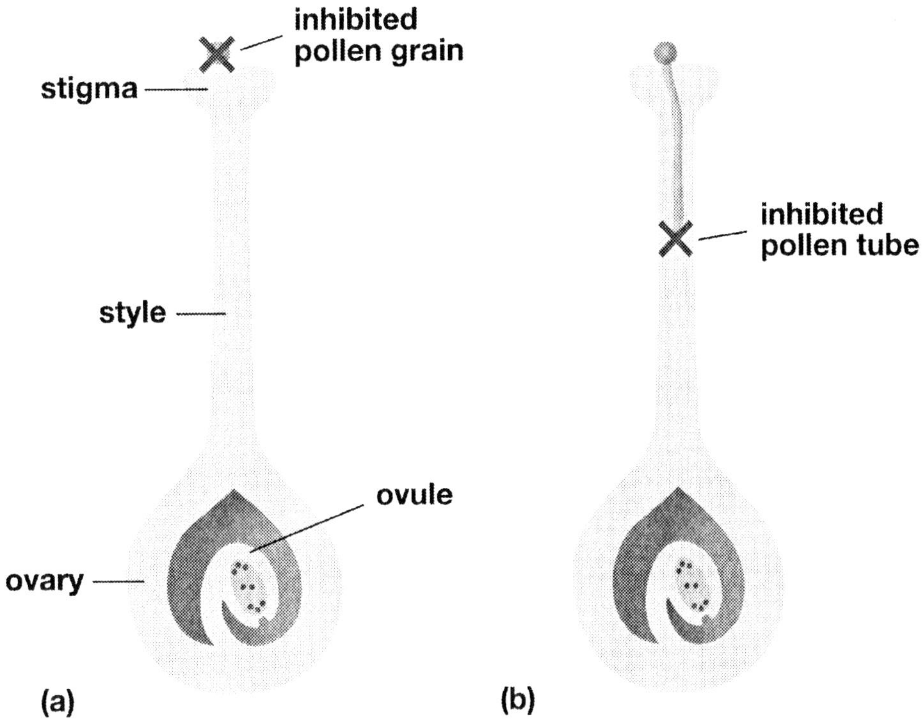

Figure 24.4 **Flower mechanisms that promote outcrossing**

NOTES

NOTES

CHAPTER 25 Principles of Ecology and the Biosphere

(a) random (b) uniform (c) clumped

Figure 25.1 **Distribution patterns of individual organisms in a population**

NOTES

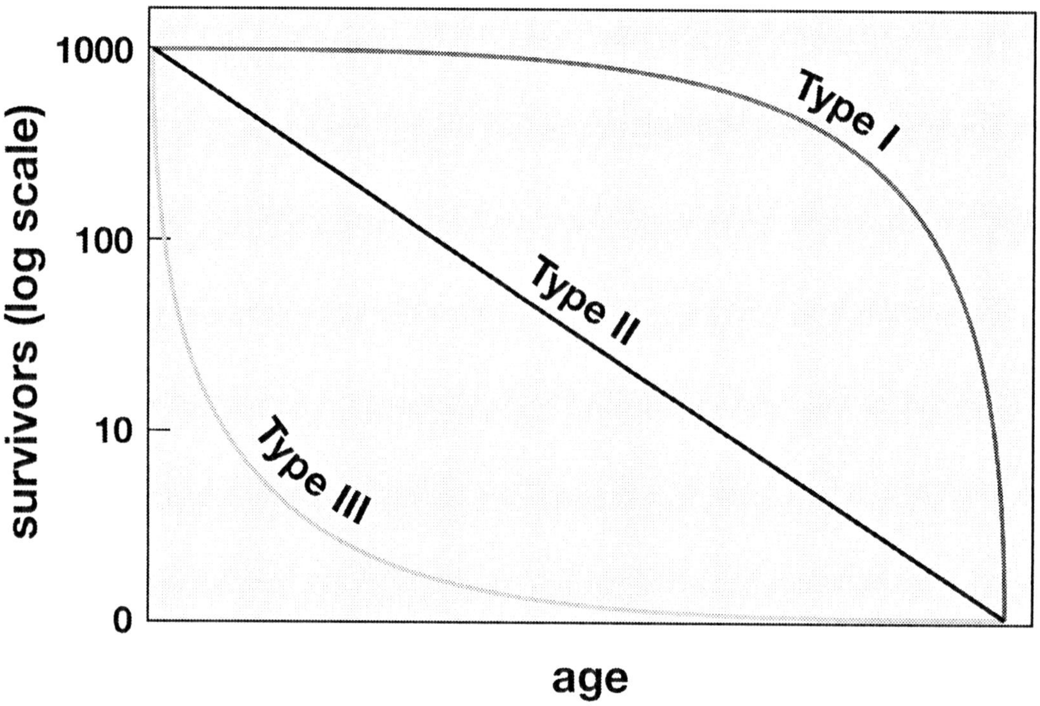

Figure 25.3 **Survivorship curves**

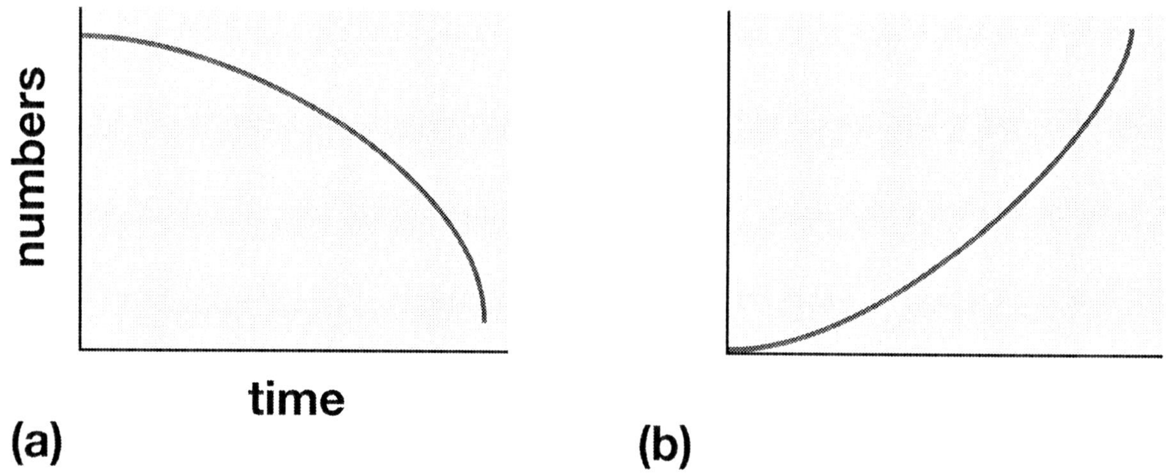

Figure 25.4 **Growth rate**

NOTES

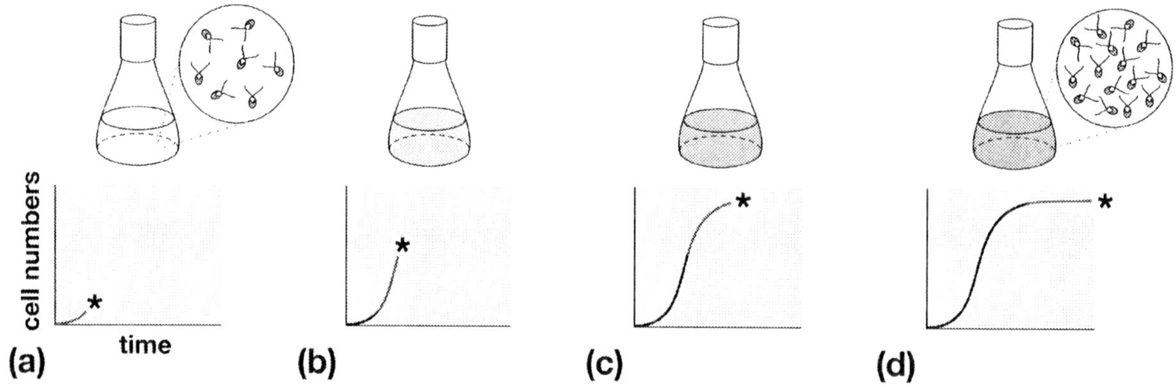

Figure 25.5 **A typical growth curve**

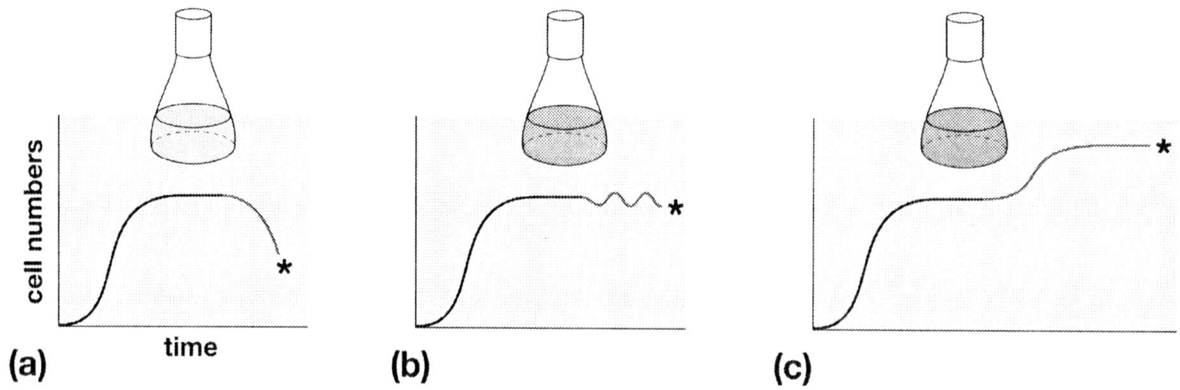

Figure 25.6 **Possible fates of population growth**

NOTES

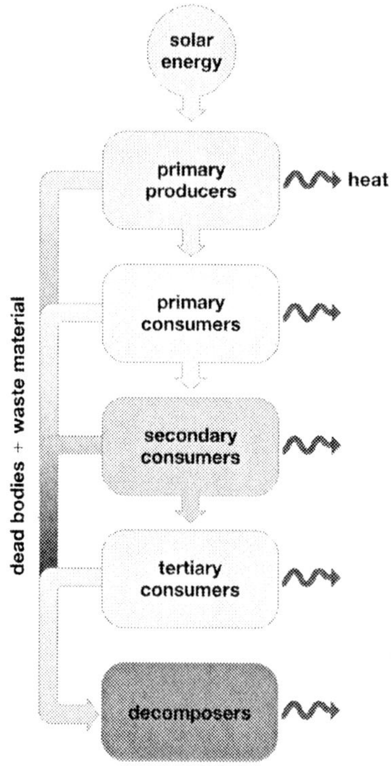

Figure 25.14 **Schematic food chain**

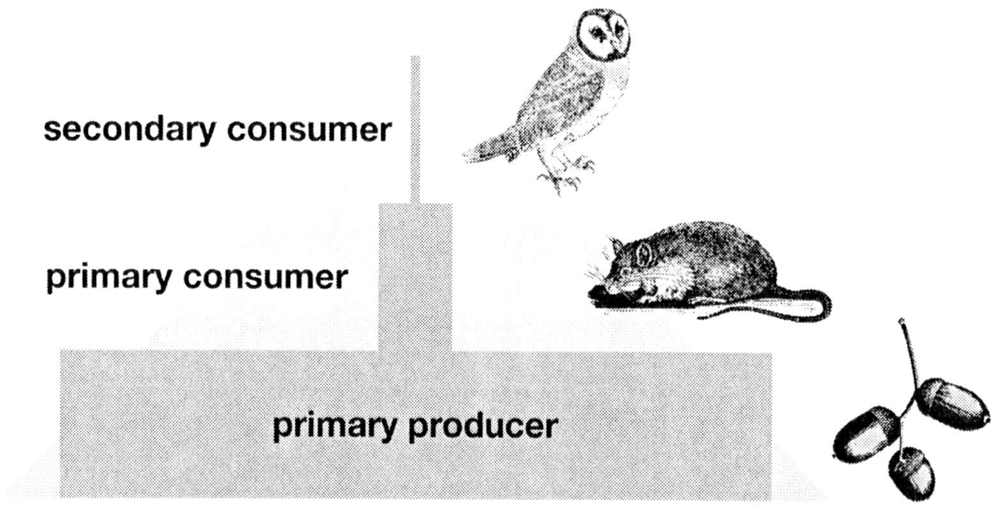

Figure 25.16 **Pyramid of biomass**

NOTES

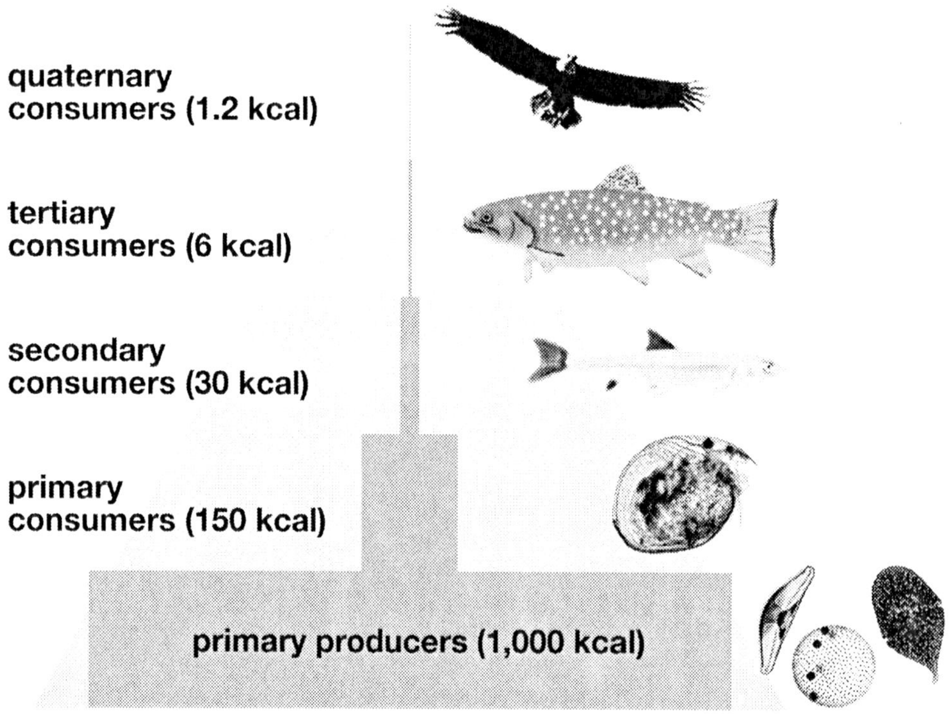

Figure 25.17 **Pyramid of energy for Lake Cayuga**

NOTES

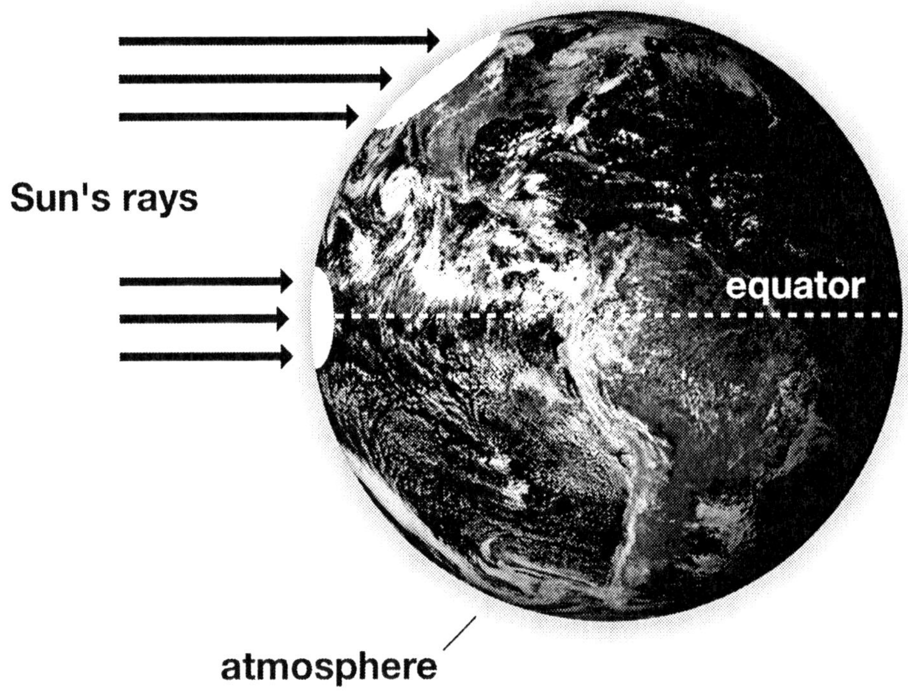

Figure 25.18 **Solar energy at the equator compared to higher latitudes**

NOTES

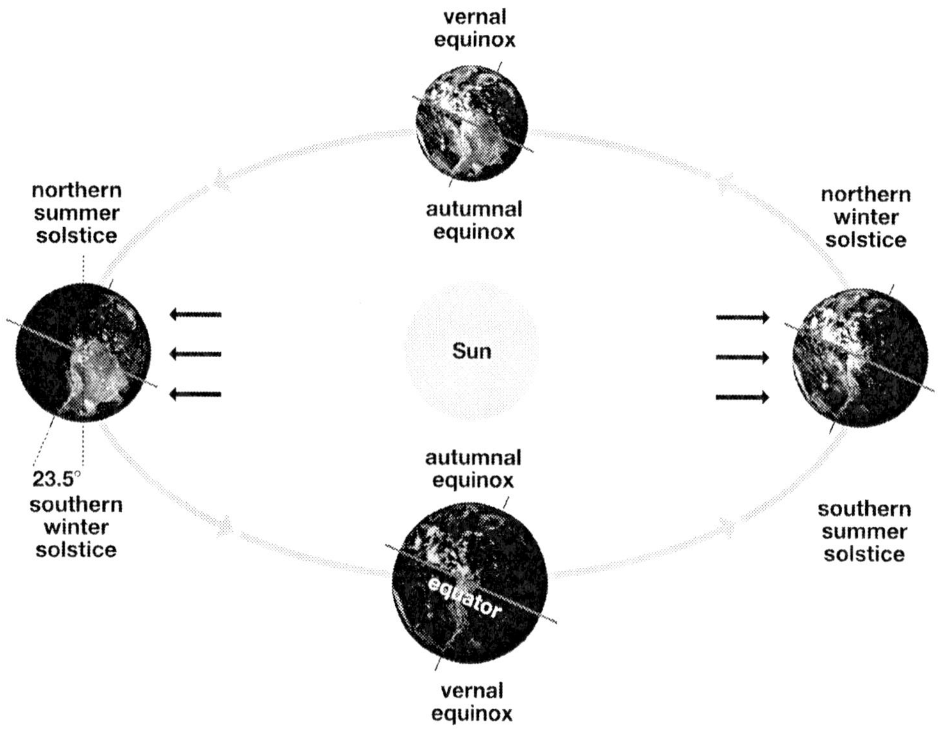

Figure 25.19 **Tilt of the Earth's axis and the seasons**

NOTES

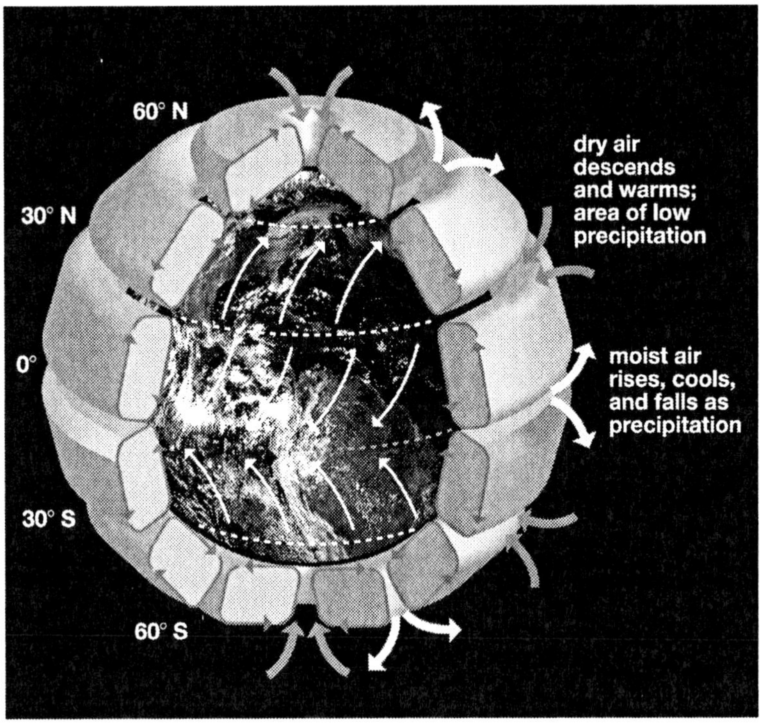

Figure 25.20 **Air cells**

NOTES

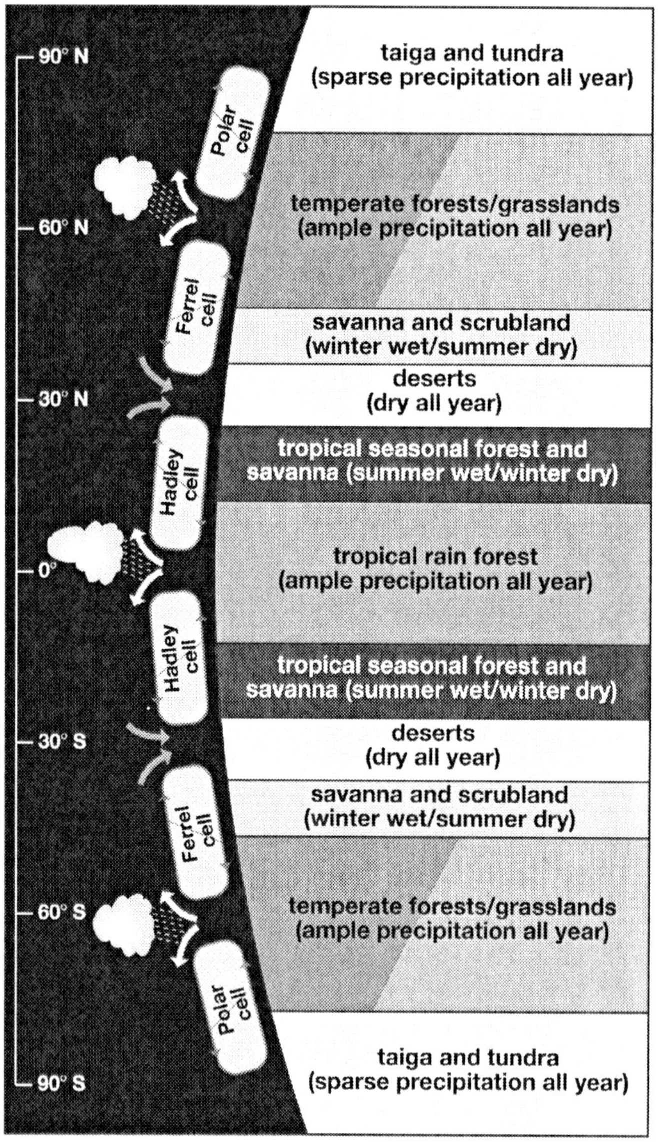

Figure 25.21 **Relationship of air cells and biomes**

NOTES

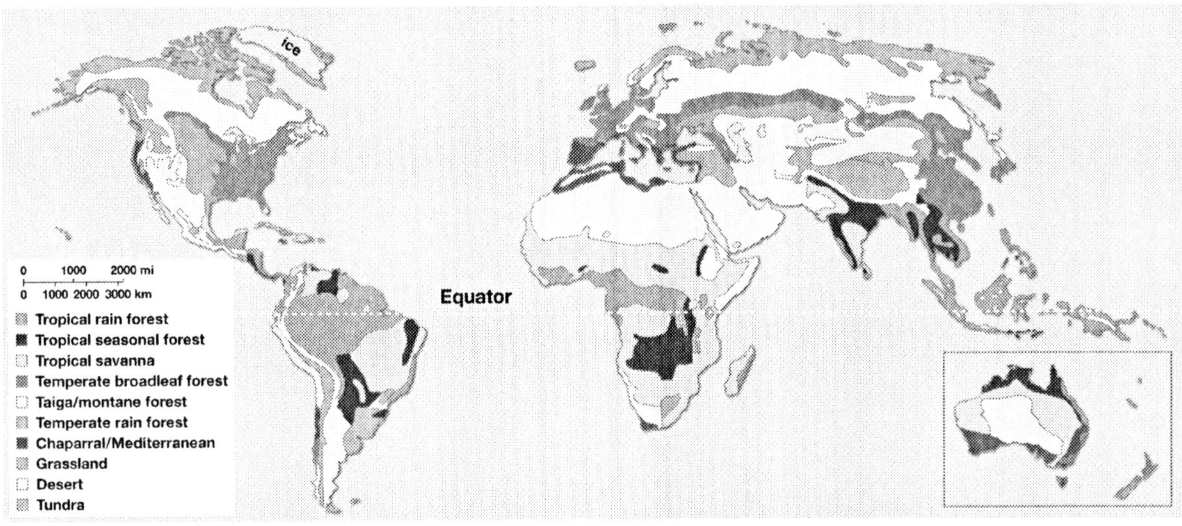

Figure 25.22 **Global distribution of the Earth's biomes**

NOTES

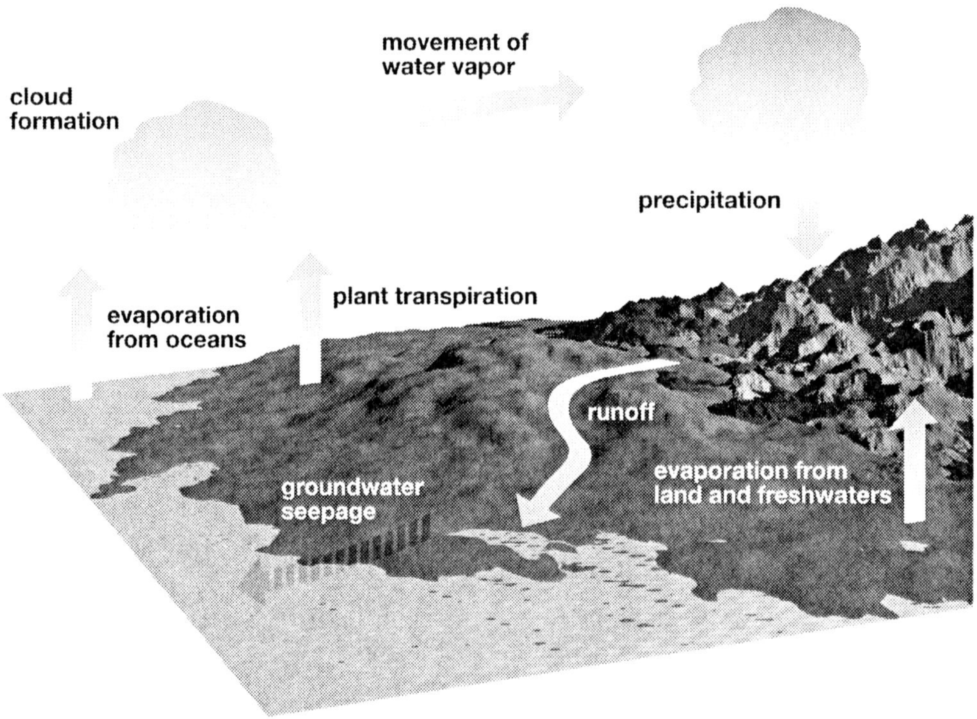

Figure 25.25 **The water cycle**

NOTES

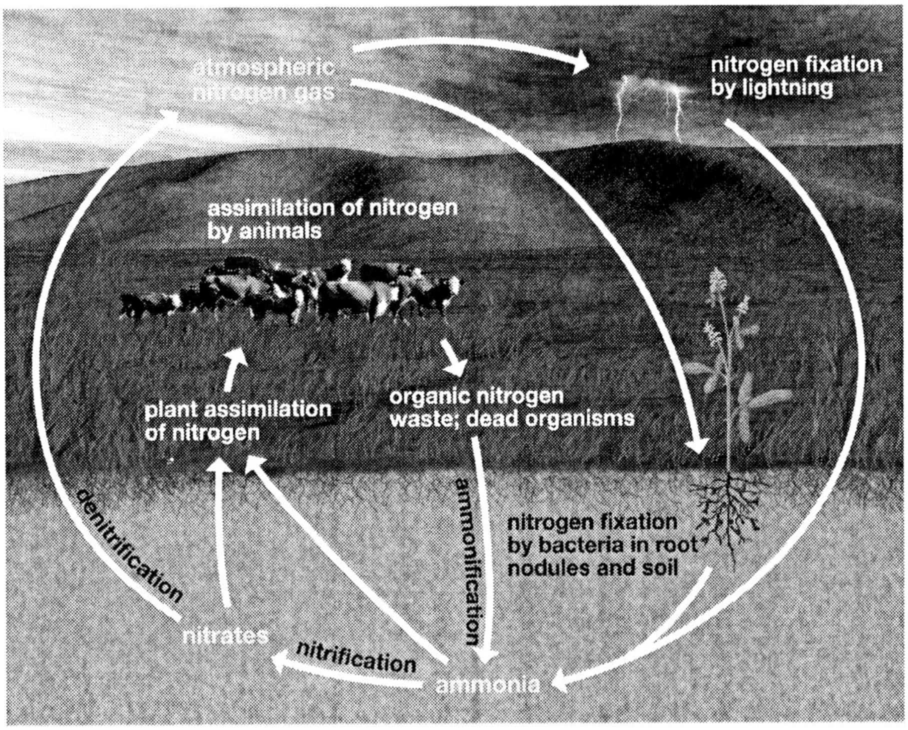

Figure 25.26 **The nitrogen cycle**

NOTES

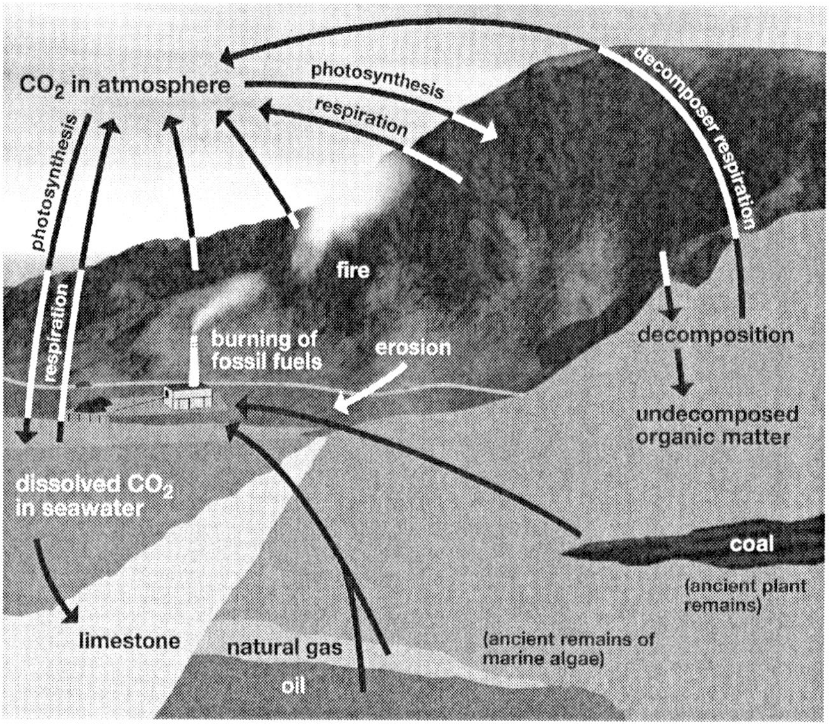

Figure 25.27 **The carbon cycle**

NOTES

CHAPTER 26

Arid Terrestrial Ecosystems

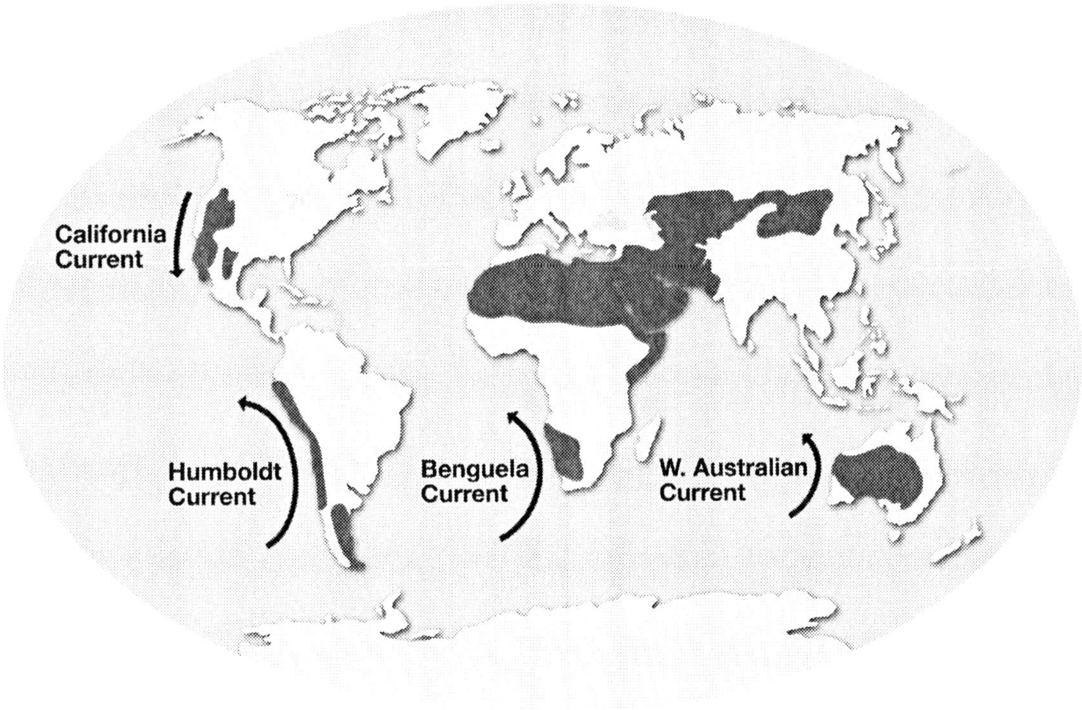

Figure 26.7 **Desert lands of the world**

NOTES

Figure 26.8 **The four American deserts**

NOTES

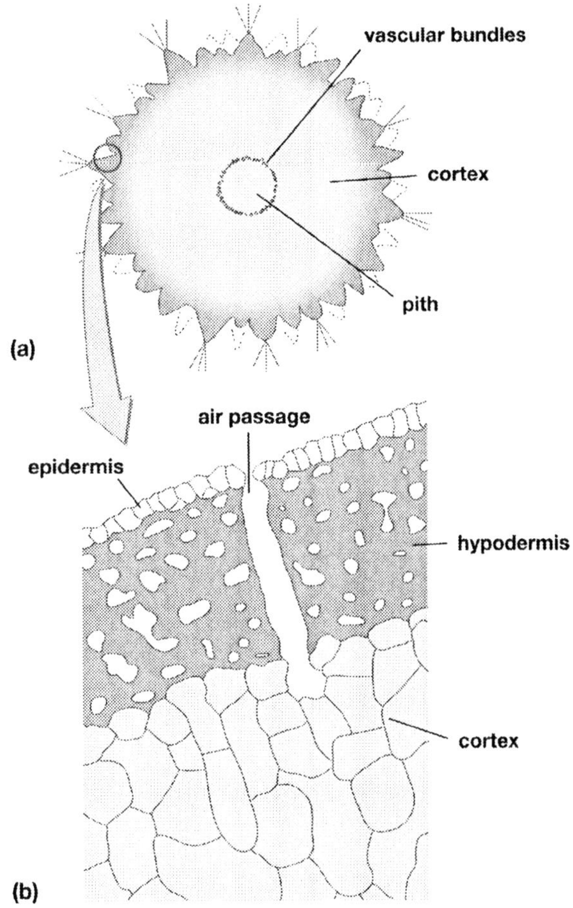

Figure 26.14 **Cactus stem anatomy**

NOTES

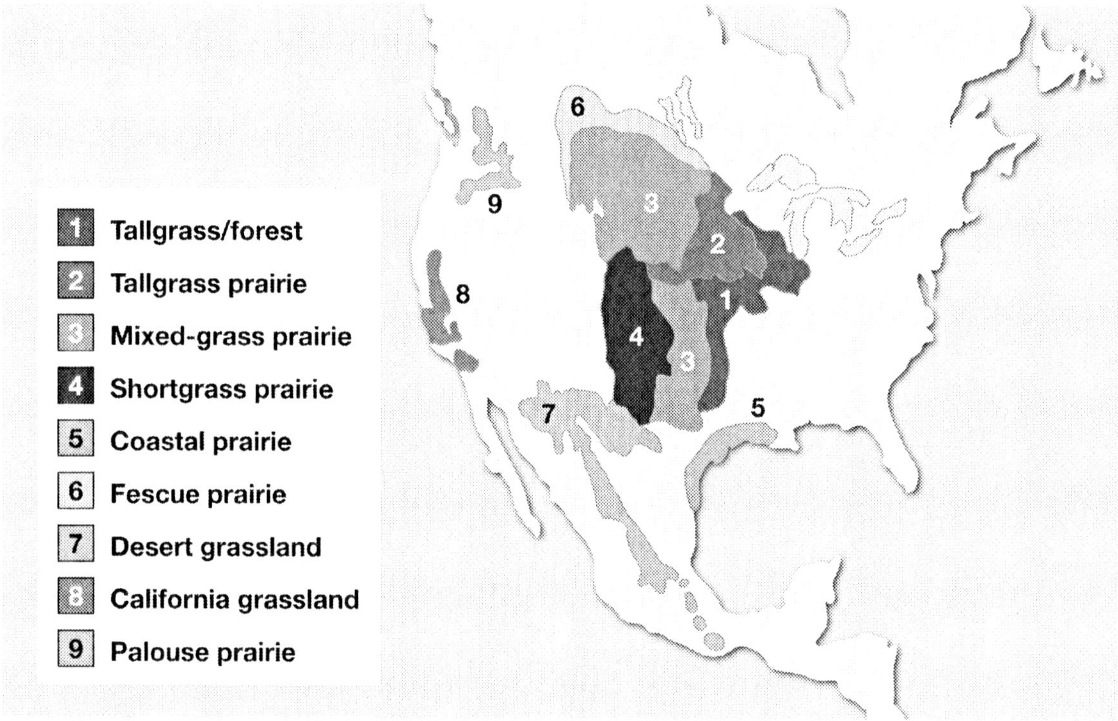

Figure 26.21 **Pre-European settlement grasslands of North America**

NOTES

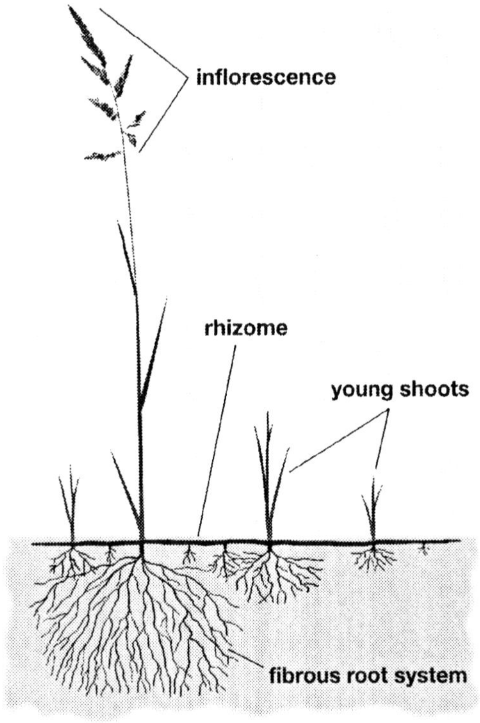

Figure 26.25 **Grass bodies include a horizontal rhizome that bears roots, leaves, and flowering plants**

NOTES

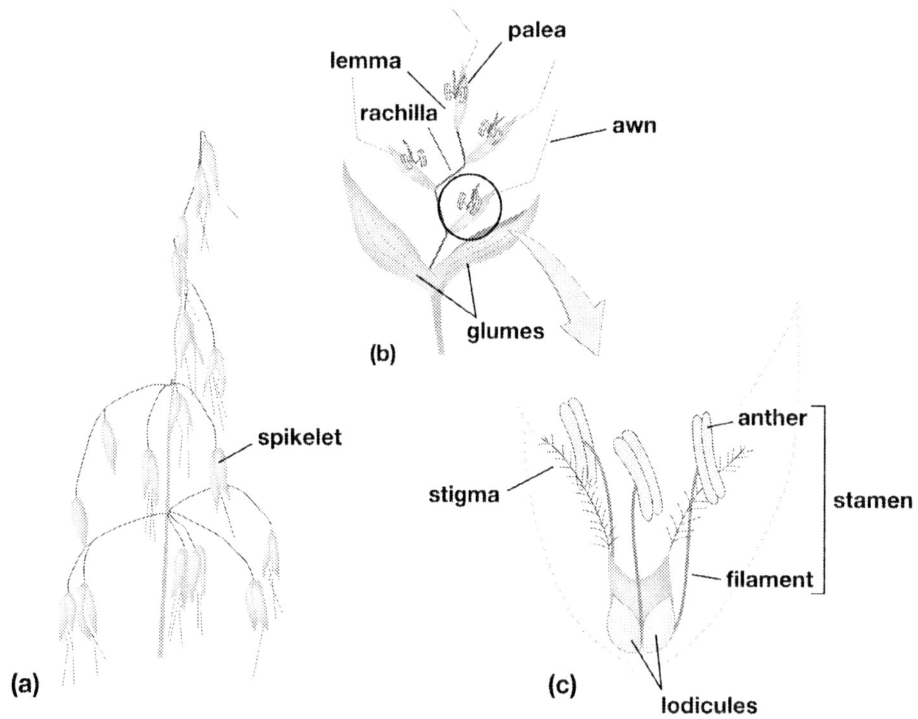

Figure 26.26 **Grass flower parts**

NOTES

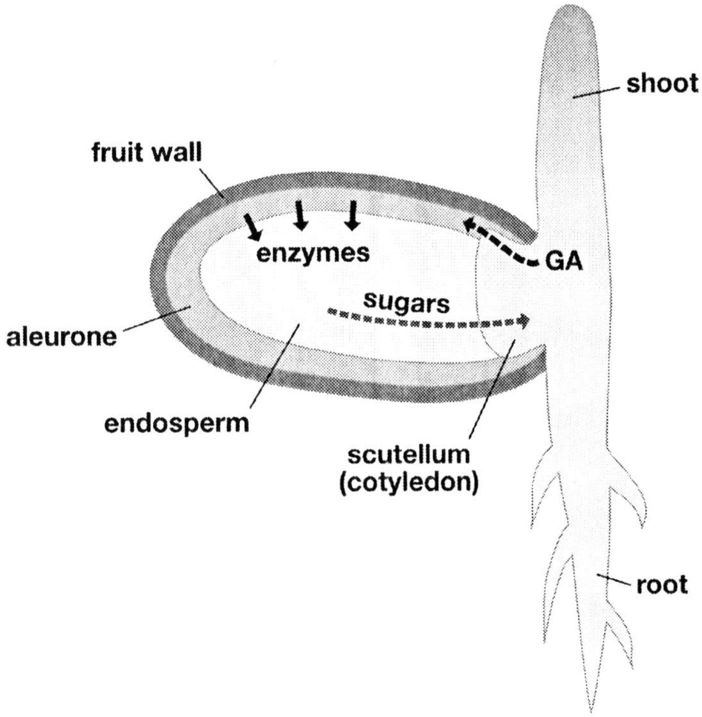

Figure 26.27 **Barley illustrates grass fruit and seed structure and function**

NOTES

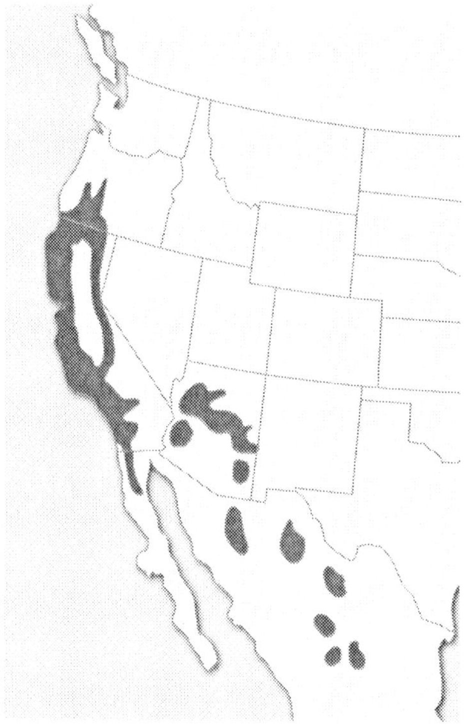

Figure 26.34 **Distribution of chaparral and similar vegetation in western North America**

NOTES

CHAPTER 27 Moist Terrestrial Ecosystems

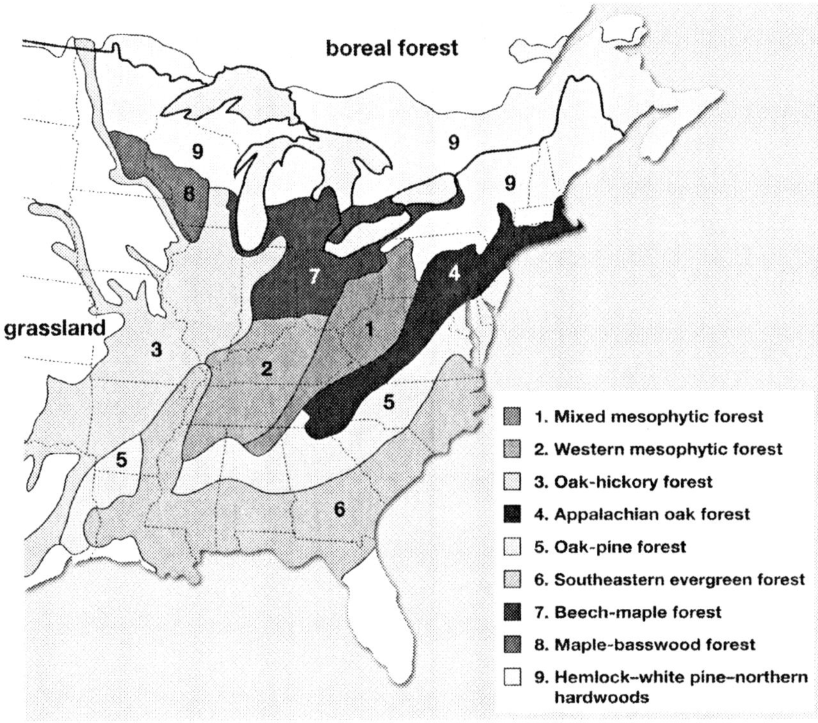

Figure 27.13 **The nine regions of eastern temperate deciduous forest**

NOTES

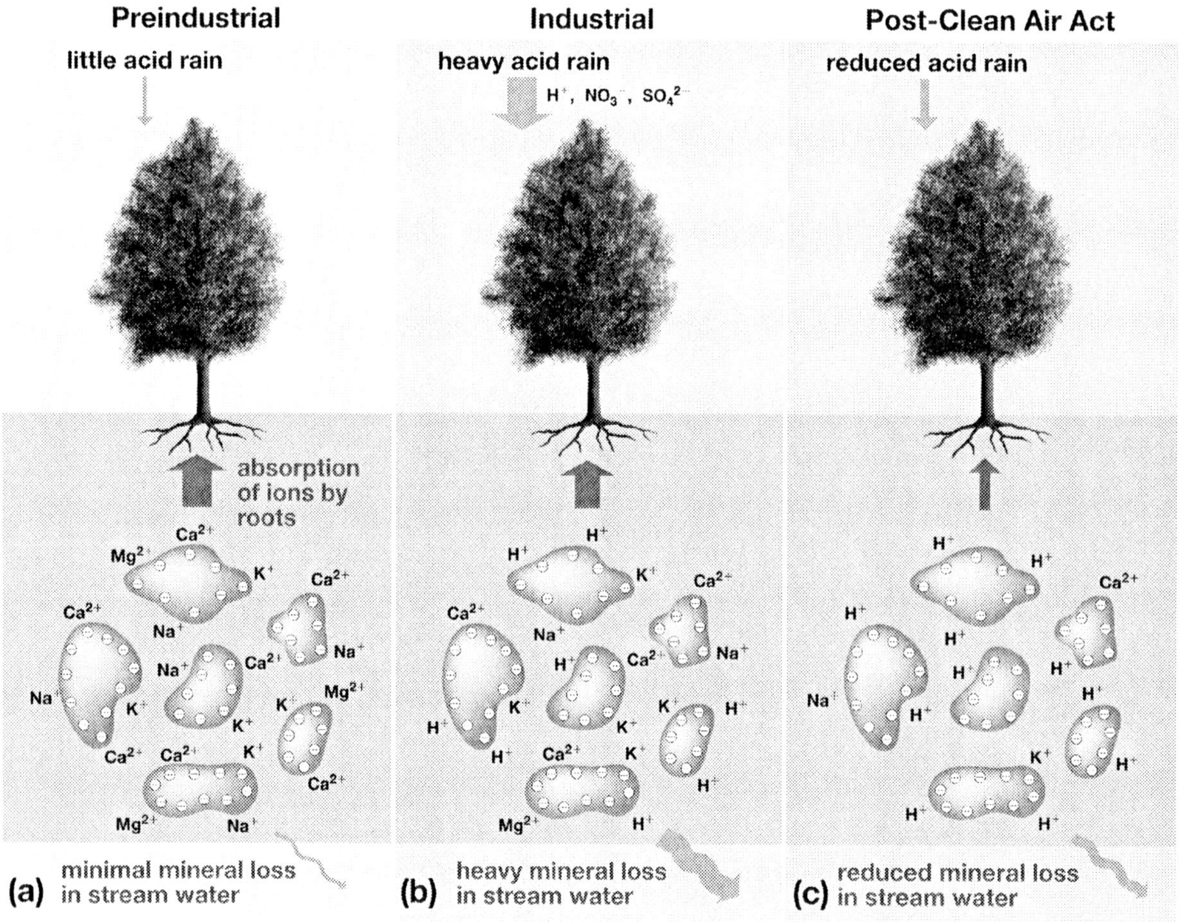

Figure 27.22 **Soil particles, mineral nutrient cations, and acid rain**

NOTES

Figure 27.29 **Closed mineral cycling in tropical forests**

NOTES

NOTES

CHAPTER 28

Aquatic Ecosystems

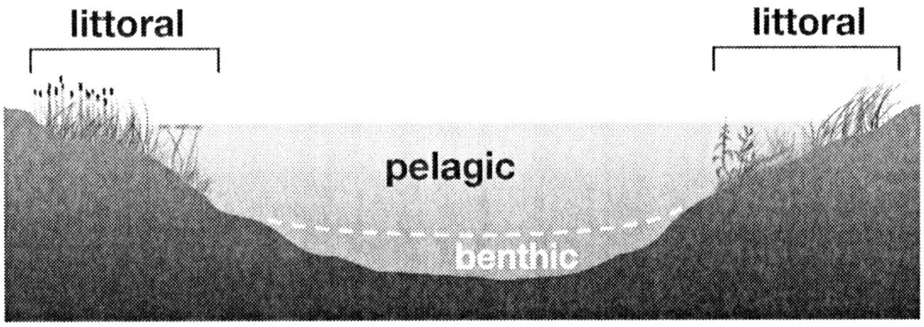

Figure 28.2 **The different habitats in a lake**

NOTES

Figure 28.3 **Temperate lakes often undergo a cycle of circulation events**

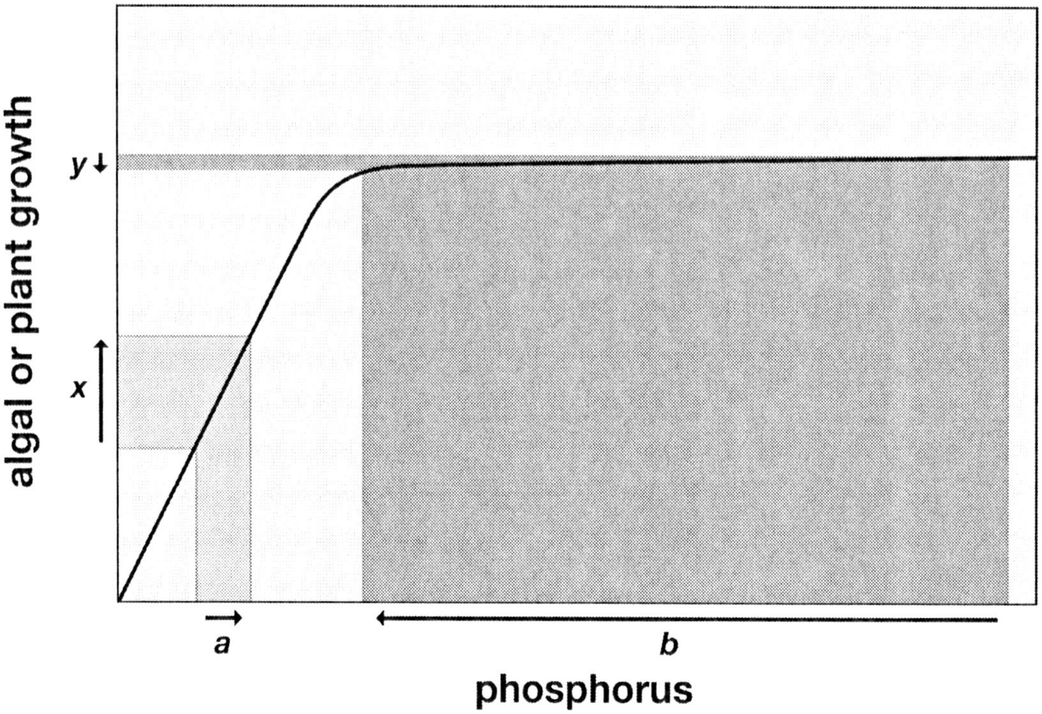

Figure 28.11 **Relationship of nutrients to algal or plant biomass**

NOTES

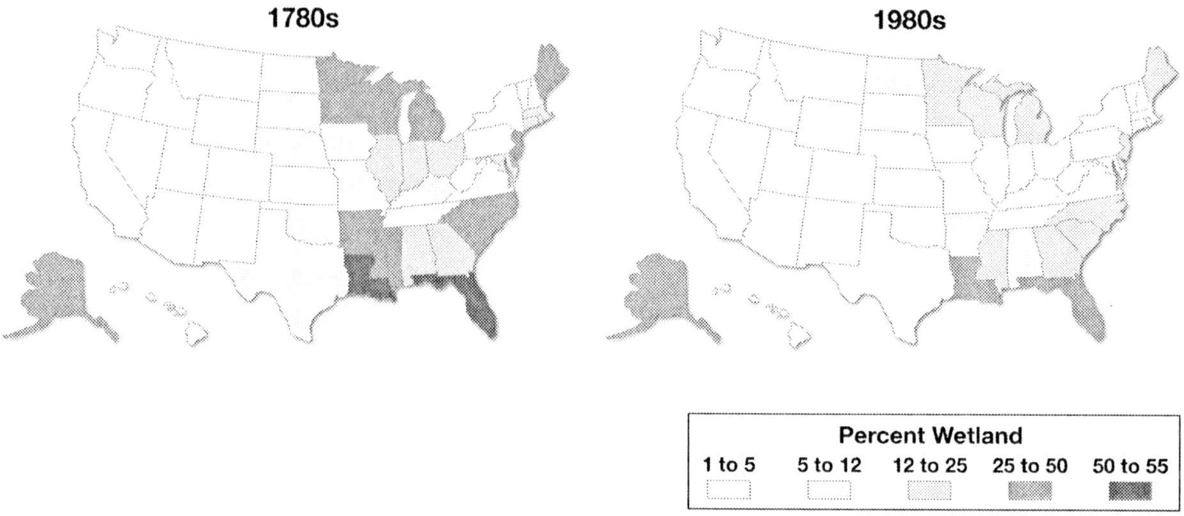

Figure 28.17 **Wetland loss**

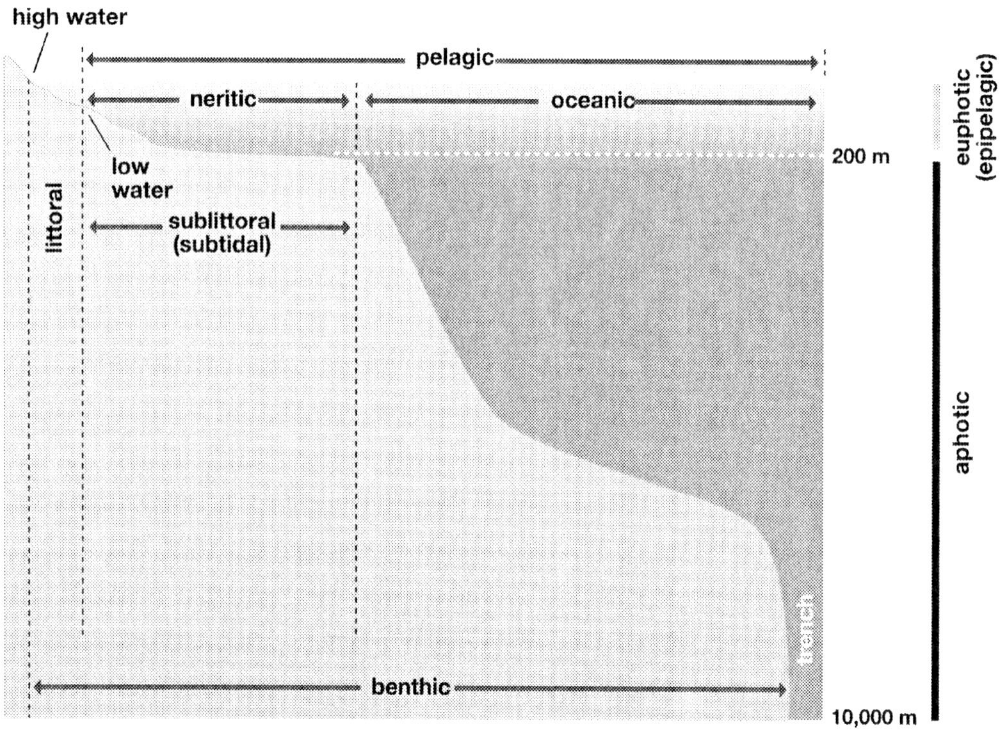

Figure 28.23 **Ocean realms**

NOTES

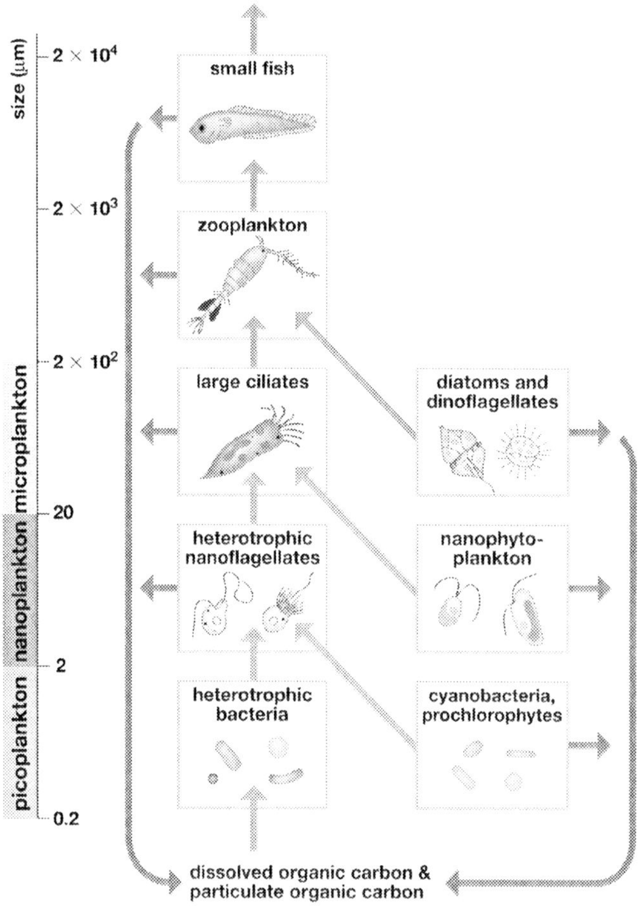

Figure 28.27 **Marine planktonic food web**

NOTES

CHAPTER 29

Human Impacts and Sustainability

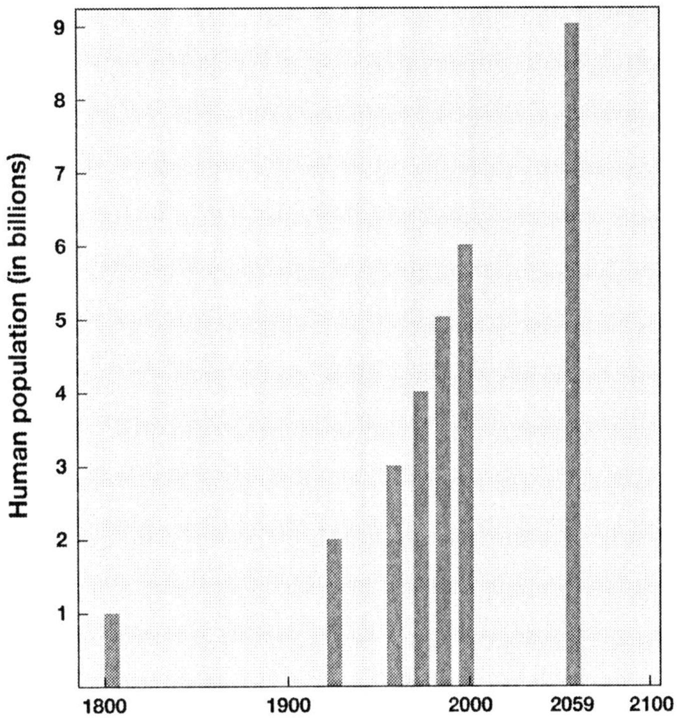

Figure 29.1 **Human population growth**

NOTES

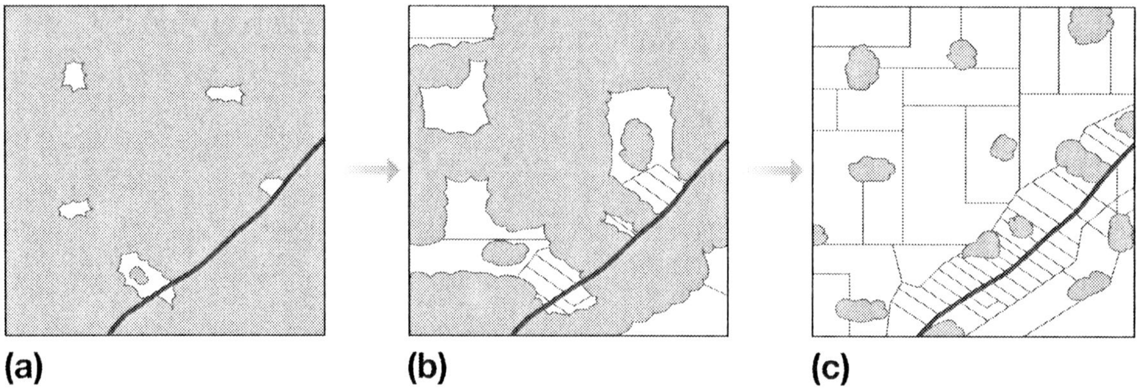

Figure 29.5 **Habitat fragmentation**

Figure 29.11 **Map of major hotspots of biodiversity in the United States**

NOTES

NOTES